Erythropoietin in the 90s

Contributions to Nephrology

Vol. 87

Series Editors
G.M. Berlyne, Brooklyn, N.Y.
S. Giovannetti, Pisa

Editorial Board
J. Churg, New York, N.Y.; *K.D.G. Edwards,* New York, N.Y.;
E.A. Friedman, Brooklyn, N.Y.; *S.G. Massry*, Los Angeles, Calif.;
R.E. Rieselbach, Milwaukee, Wisc.; *J. Traeger*, Lyon

KARGER

Basel · München · Paris · London · NewYork · New Delhi · Bangkok · Singapore · Tokyo · Sydney

International Symposium, Würzburg, March 23–24, 1990

Erythropoietin in the 90s

Volume Editors
R.M. Schaefer, A. Heidland, Würzburg
W.H. Hörl, Homburg/Saar

36 figures and 20 tables, 1990

Basel · München · Paris · London · NewYork · New Delhi · Bangkok · Singapore · Tokyo · Sydney

Contributions to Nephrology

Library of Congress Cataloging-in-Publication Data
　　Erythropoietin in the 90s / volume editors, R.M. Schaefer, A. Heidland, W.H. Hörl.
　　(Contributions to nephrology; vol 87)
　　International symposium, Würzburg, March 23-24, 1990.
　　Includes bibliographical references.
　　1. Renal anemia – Chemotherapy – Congresses.　2. Recombinant erythropoietin –
　　Therapeutic use – Congresses.　3. Erythropoietin – Congresses.　4. Hemodialysis – Congresses.
　　I. Schaefer, R.M.　II. Heidland, August.　III. Hörl, Walter H.　IV. Series:
　　Contributions to nephrology: v. 87.
　　[DNLM: 1. Erythropoietin – physiology – congresses. 2. Erythropoietin – therapeutic use –
　　congresses.　W1 CO778UN v. 87 / WH 150 E739 1990]
　　RC641.7.R44E792 1990
　　616.6' 14061 – dc20
　　ISBN 3-8055-5267-X

Bibliographic Indices
　　This publication is listed in bibliographic services, including Current Contents® and Index Medicus.

Drug Dosage
　　The authors and the publisher have exerted every effort to ensure that drug selection and dosage set forth in this text are in accord with current recommendations and practice at the time of publication. However, in view of ongoing research, changes in government regulations, and the constant flow of information relating to drug therapy and drug reactions, the reader is urged to check the package insert for each drug for any change in indications and dosage and for added warnings and precautions. This is particularly important when the recommended agent is a new and/or infrequently employed drug.

All rights reserved.
　　No part of this publication may be translated into other languages, reproduced or utilized in any form or by any means, electronic or mechanical, including photocopying, recording, microcopying, or by any information storage and retrieval system, without permission in writing from the publisher.

© 　Copyright 1990 by S. Karger AG, P.O. Box, CH-4009 Basel (Switzerland)
　　ISBN 3-8055-5267-X

Contents

Preface VII

Geissler, K.; Stockenhuber, F.; Hinterberger, W.; Balcke, B.; Lechner, K. (Wien): Recombinant Human Erythropoietin: A Multipotential Hemopoietic Growth Factor in vivo and in vitro 1

Hanicki Z.; Sułowicz, W.; Stepniewski, M.; Kuźniewski, M.; Kraśniak, A.; Kopeć, J.; Cieszkowska, U.; Stolarska, K. (Kraków): Recombinant Human Erythropoietin Stimulates Synthesis of Fetal Haemoglobin in Haemodialysed Patients with Anaemia due to End-Stage Kidney 11

Kurtz, A.; Eckardt, K.-U. (Zürich): Erythropoietin Production in Chronic Renal Disease before and after Transplantation 15

Oster, W. (Marburg); Mertelsmann, R. (Freiburg i. Br.): The Role of Erythropoietin in Patients with Anemia and Normal Renal Function 26

Paczek, L.; Schaefer, R.M.; Heidland, A. (Würzburg): Improved Function of B Lymphocytes in Dialysis Patients Treated by Recombinant Human Erythropoietin 36

Kokot, F.; Więcek, A.; Grezeszczak, W.; Klin, M.; Żukowska-Szczechowska, E. (Katowice): Influence of Erythropoietin Treatment on Glucose Tolerance, Insulin, Glucagon, Gastrin and Pancreatic Polypeptide Secretion in Haemodialyzed Patients with End-Stage Renal Failure 42

Schmidt, R. (Rostock); Lerche, D. (Berlin); Dörp, E.; Winkler, R.; Klinkmann, H. (Rostock): Changes in Red Blood Cell Volume under Recombinant Human Erythropoietin Therapy 52

Bommer, J.; Barth, H.P.; Schwöbel, B. (Heidelberg): rhEPO Treatment of Anemia in Uremic Patients 59

Jelkmann, W.; Wolff, M. (Bonn); Fandrey, J. (Lübeck): Modulation of the Production of Erythropoietin by Cytokines: In vitro Studies and Their Clinical Implications 68

Hörl, W.H.; Dreyling, K.; Steinhauer, H.B.; Engelhardt, R.; Schollmeyer, P. (Freiburg i. Br.): Iron Status of Dialysis Patients under rhuEPO Therapy .. 78

Casati, S.; Campise, M.; Ponticelli, C. (Milano): Aluminium Interference in the Treatment with Recombinant Human Erythropoietin 87

Contents

Schollmeyer, P.; Lubrich-Birkner, I.; Steinhauer, H.B. (Freiburg i. Br.): Effect of Recombinant Human Erythropoietin on Anemia and Dialysis: Efficiency in Patients Undergoing CAPD 95
Koene, R.A.P.; Frenken, L.A.M. (Nijmegen): Does Treatment of Predialysis Patients with Recombinant Human Erythropoietin Compromise Renal Function? . . . 105

Subject Index 113

Preface

The present volume of *Contributions to Nephrology* reports on a workshop entitled Erythropoietin in the 90s, which was held in March 1990 in Würzburg. As indicated by this title, the main intention was to address major issues arising during routine therapy in hemodialysis patients and to provide an outlook into areas of future applications of erythropoietin. Accordingly, there were reports on cardiovascular changes, aluminum interference, iron substitution, and route of administration in hemodialysis patients. Two presentations dealt with erythropoietin in predialysis and CAPD patients. One paper addressed the use of erythropoietin in the anemia of malignancy. A couple of reports dealt with effects on cells other than red progenitors, such as B lymphocytes and non-red progenitor cells. In addition, effects on hemoglobin F synthesis and endocrine alterations during erythropoietin were also described. Finally, there was one article on serum erythropoietin levels after renal grafting and one report on the influence of cytokines on erythropoietin production by the HepG2 cell line. It can be expected that this monograph will prove useful not only to the nephrologist, but also to investigators working in the fields of physiology, endocrinology, and oncology. This meeting would not have been possible without the generous support of Cilag GmbH, Sulzbach/Ts.

R.M. Schaefer

Recombinant Human Erythropoietin: A Multipotential Hemopoietic Growth Factor in vivo and in vitro

K. Geissler[a], F. Stockenhuber[b], W. Hinterberger[a], B. Balcke[b], K. Lechner[a]

Abteilungen für [a]Hämatologie und [b]Nephrologie,
1. Medizinische Universitätsklinik, Wien, Österreich

The first clinical trials of recombinant hemopoietic growth factors in man indicate that a new form of medical therapy has evolved. Erythropoietin (EPO) was the first clearly defined humoral biomolecule regulating hemopoiesis [1] and it was also the first hemopoietic growth factor which was used for therapeutic purposes in recombinant form [2]. It is well established that EPO is the major humoral regulator of red cell mass. It is a circulating glycoprotein hormone that is produced largely by the kidney [3], although 5–10% appear to be derived from an extrarenal source, possibly the liver. EPO was found to act at the progenitor cell level, in vitro it stimulates erythroid progenitor cells to induce both proliferation and differentiation of erythroid cells [4, 5].

Administration of recombinant human erythropoietin (rhEPO) to patients with end-stage renal failure has proved to be highly effective in correcting the anemia of advanced renal disease. Several authors have described a dose-dependent increase of hematocrit level and reticulocyte counts under treatment with rhEPO [2, 6, 7] but there are only few data about the impact of rhEPO on hemopoietic progenitor cells in vivo.

The aim of the present study was to analyze the effect of rhEPO treatment on the peripheral blood (PB) progenitor cell compartment including myeloid (CFU-GM), erythroid (BFU-E) and multipotential (CFU-MIX) progenitor cells.

Table 1. Baseline data collected during the run-in period: patients treated with 50 or 80 U rhEPO/kg b.w. 3 times/week

	G 50		G 80	
Number of patients	9		11	
Age, years	47	(21–67)	48	(27–71)
Duration of dialysis, months	36	(7–107)	39	(5–89)
Hematocrit, %	21.1	(18–24)	21.0	(17–25)
Corrected reticulocyte, %	0.67	(0.3–1.7)	0.65	(0.2–1.9)

Materials and Methods

Patients

Twenty patients receiving chronic hemodialysis entered the study to be treated with rhEPO (Boehringer Mannheim, Mannheim, FRG). After informed consent had been obtained patients were randomly assigned to receive either 50 U/kg b.w. (group G 50; n = 9) or 80 U/kg b.w. (group G 80; n = 11) intravenously 3 times weekly after each hemodialysis session.

Uremia resulted from glomerulonephritis (n = 8), pyelonephritis (n = 6), analgetic abuse (n = 4), and polycystic kidney disease (n = 1). In 1 patient the underlying disease was unknown. Baseline data of patients enrolled in the study are shown in table 1. None of the patients had received immunosuppressive therapy for the last 3 months before rhEPO treatment. There were no symptoms or clinical findings of vitamin B_{12}, iron or folic acid deficiency, nor hyperparathyroidism or intoxication of aluminum.

Recombinant Human Erythropoietin

The human EPO used in this study was obtained by DNA technology and has been developed jointly by Genetics Institute, Cambridge, Mass., USA, and Boehringer Mannheim, FRG [8]. rhEPO was more than 98% pure, and was formulated in a buffered saline solution containing 1% bovine serum albumin. The specific activity of rhEPO is 173,000 U/mg hormone.

Laboratory Measurements

Hematological profile (Coulter counter; Coulter Electronics, Hialeah, Fla., USA) was measured before each dialysis session. Blood chemistry was determined once a week by a multichannel autoanalyzer. Total protein, electrophoresis, acid base status and blood coagulation parameters were controlled once a month. Ferritin concentration was detected by radioimmunoassay and controlled monthly. Reticulocyte counts were corrected for the degree of anemia.

Cell Separation Procedure

Twenty milliliters of PB were collected in 4 ml EDTA under sterile conditions. The light density mononuclear cells (MNC) were obtained from the interface after sedimentation over Ficoll-Hypaque (400 *g*, 40 min, density 1.077 g/ml).

In vitro Assay

The hemopoietic progenitor cells in PB were determined weekly. Blood samples were taken 48 h after each injection of rhEPO. Multipotent (CFU-MIX) as well as committed progenitor cells (BFU-E, CFU-GM) were assayed using a modification of the clonal assay described by Fauser and Messner [9]. Each plate contained 0.9% methylcellulose, 30% fetal calf serum, 10% bovine serum albumin (Behring, FRG), 1 U/ml EPO (Toyobo), alpha-thioglycerol (10^{-4} mol/l), 5% phytohemaglutinin-leukocyte-conditioned medium and Iscove's modified Dulbecco's medium (Gibco, Grand Island, N.Y., USA). Peripheral blood mononuclear cells were plated in quadruplicates at 2.5×10^5/ml. After a culture period of 14 days (at 37 °C with 5% CO_2 and full humidity) cultures were examined under an inverted microscope. Aggregates with at least 40 translucent, dispersed cells were counted as CFU-GM. Bursts containing more than 100 red-colored cells were scored as BFU-E. CFU-MIX were identified by their heterogeneous composition of translucent and hemoglobinized cells. Individual colonies suspected as CFU-MIX were picked, transferred to glass and stained by May-Grünwald-Giemsa for cytological examination under a light microscope. The clonal nature of mixed colonies was confirmed by the linear relationship between the numbers of colonies.

Calculation of the Peripheral Blood Progenitors

The total number of mononuclear cells per milliliter of peripheral blood was determined by multiplying the total number of white cells by the percentage of mononuclear cells in the differential count. The mononuclear cell fraction consisted of immature myeloid forms, monocytes and lymphocytes circulating in the peripheral blood. The total numbers of progenitor cells per milliliter of peripheral blood were determined by multiplying the number of CFU-GM, BFU-E or CFU-MIX/10^5 MNC by the total number of MNC/ml of PB. The normal ranges for the hemopoietic progenitor cells were obtained from 20 healthy volunteers.

Statistical Analysis

Statistical analysis was carried out using the Mann-Whitney U test. The level of significance was set at $p < 0.05$. All values are expressed as means and standard error.

Results

Since none of the patients had their therapy discontinued, data for evaluation could be obtained from all 20 patients. The frequency of dialysis sessions remained constant in all patients. The injection of rhEPO was well tolerated in all cases without any problems of local or systemic intolerance. Mild side effects were observed in 8 patients. Worsening of hypertension that needed increase of antihypertensive treatment was detected in 2 patients of group G 50 and in 3 patients in group G 80. One patient from each group complained about headache and pain in the joints. Influenza-like symptoms occurred in 1 patient of group G 80 but were transitory.

Hematological Findings and Iron Status

Reticulocyte count started to rise within the second week in both groups, and was followed by an increase of hematocrit during the third week in group G 80 and after 4 weeks of rhEPO treatment in group G 50. The median percent increment of hematocrit per week was 1% in group G 50 and 1.2% in group G 80. No significant change of mean corpuscular volume (MCV) could be detected. Platelet number increased moderately under rhEPO therapy by an average of 7% in both groups, but true thrombocytosis was not observed in any of the patients. The increase of platelet count was not accompanied by any complication, such as thrombosis of the arteriovenous fistulas or other thromboembolic events. The increase of thrombocyte count was not intercorrelated with iron deficiency. White blood counts (WBC) and differential counts remained unaffected. Along with the increase of hematocrit, a considerable decrease of serum iron and ferritin levels was observed, necessitating iron supplementation in 4 patients of G 50 and 5 patients in group G 80.

Blood and Serum Chemistry

There was no increase of potassium level during EPO therapy. The median serum concentration of urea and creatinine did not change significantly. Furthermore, no change was found for bilirubin, liver enzymes or lactate dehydrogenase (LDH). Serum protein concentration and electrophoresis remained constant.

Peripheral Progenitor Cells

The concentration of hemopoietic progenitor cells BFU-E in PB increased significantly from subnormal levels to normal range within 1 week of treatment in both treatment groups. The rise was most pronounced during the first week and the concentration remained within normal ranges up to the end of the observation period (fig. 1).

The number of CFU-MIX showed a similar course with a marked increase from subnormal ranges within 1 week of the two rhEPO supplementation regimens (fig. 2). The number of CFU-GM proved to be low but within normal ranges prior to rhEPO therapy. rhEPO therapy led to a significant rise of CFU-GM levels in both groups (fig. 3).

One patient acquired bacterial pneumonia during rhEPO treatment. The onset of clinical symptoms was accompanied by a marked decrease of BFU-E levels which was followed by a reduction of hematocrit with a delay of 2 weeks, although therapy with rhEPO was continued. Antibiotic therapy was started and the patient recovered from respiratory infection within

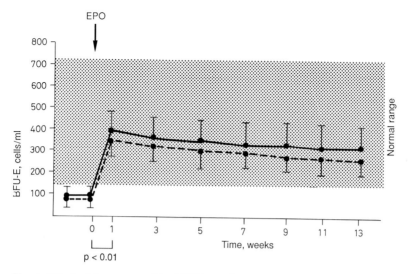

Fig. 1. Effect of treatment with rhEPO on the number of erythroid progenitors BFU-E in the peripheral blood of patients with anemia of chronic renal failure. —— = 80 U/kg b.w. 3 times/week (n = 11); --- = 50 U/kg b.w. 3 times/week (n = 9).

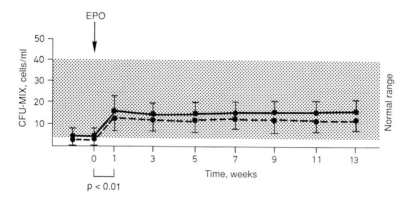

Fig. 2. Effect of treatment with rhEPO on the number of multipotential hemopoietic progenitors CFU-MIX in the peripheral blood of patients with anemia of chronic renal failure. —— = 80 U/kg b.w. 3 times/week (n = 11); --- = 50 U/kg b.w. 3 times/week (n = 9).

20 days. The counts of BFU-E as well as the red blood count returned to levels measured before infection after a further 2 and 4 weeks, respectively (fig. 4). Patients who developed iron deficiency showed a decrease of BFU-E levels 1 week prior to a marked fall of hematocrit, as shown for 1 patient in figure 5.

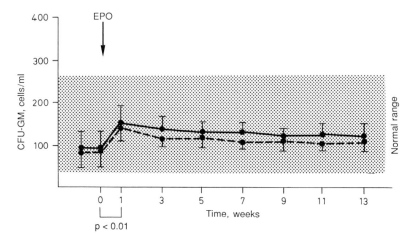

Fig. 3. Effect of treatment with rhEPO on the number of granulocyte-macrophage progenitors CFU-GM in the peripheral blood of patients with anemia of chronic renal failure. ──── = 80 U/kg b.w. 3 times/week (n = 11); ─ ─ ─ = 50 U/kg b.w. 3 times/week (n = 9).

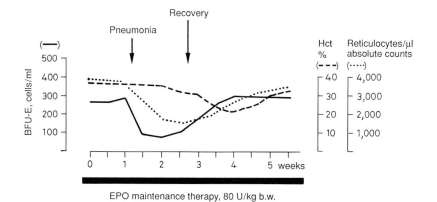

Fig. 4. Effect of a bacterial infection on the number of erythroid progenitors BFU-E in the peripheral blood of a patient with anemia of chronic renal failure during continuous rhEPO treatment. ──── = BFU-E levels; ─ ─ ─ = hematocrit; · · · = absolute count of reticulocytes.

Effect of rhEPO in vitro

If rhEPO was added to bone marrow mononuclear cells (BM MNC) from a normal donor in vitro, it markedly stimulated growth of BFU-E but had little or no effect on growth of CFU-GM and CFU-MIX (table 2). In contrast, rhEPO not only increased number and size of BFU-E but surprisingly also

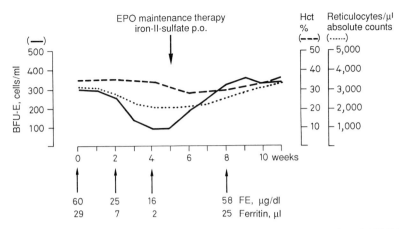

Fig. 5. Effect of an iron deficiency on the number of erythroid progenitors BFU-E in the peripheral blood of a patient with anemia of chronic renal failure during continuous rhEPO treatment. —— = BFU-E levels; --- = hematocrit; ··· = absolute count of reticulocytes.

Table 2. Effect of rhEPO on the in vitro growth of CFU-GM, BFU-E and CFU-MIX from normal bone marrow cells with or without pretreatment with rhIL-3 (one of three representative experiments)

3 h preincubation	14 days' incubation	Number of colonies/30,000 BM MNC ± SD		
		CFU-GM	BFU-E	CFU-MIX
100 U/ml rhIL-3	medium	8.5 ± 0.7	0.0 ± 0.0	0.0 ± 0.0
100 U/ml rhIL-3	2 U/ml rhEPO	44.0 ± 5.7	23.5 ± 0.7	1.0 ± 0.0
Medium	2 U/ml rhEPO	2.0 ± 1.4	13.0 ± 4.2	0.0 ± 0.0

stimulated CFU-GM and CFU-MIX if BM MNC were preincubated with rhIL-3 for 3 h. The majority of myeloid colonies stimulated by rhEPO from rhIL-3 preincubated BM MNC consisted of monocyte/macrophage cells.

Discussion

In agreement with the findings reported by others [2, 6, 7], we found that administration of rhEPO in a dosage of 50 or 80 U/kg b.w. 3 times weekly exerts an effective stimulus on red cell production in patients on regular hemodialysis. An increase of mean hematocrit to 30% and mean hemoglobin concentration to 10.9 g/dl was achieved within 10 weeks of treatment in both groups.

The principal target cells of EPO are erythroid progenitor cells. Thus EPO acts on the CFU-E, the immediate precursor of the differentiated normoblasts, and to a lesser extent on the BFU-E, the more immature precursor [4]. A stimulatory activity of EPO on BFU-E in tissue culture has been reported [5] and EPO has induced DNA synthesis in human marrow BFU-E in vitro [10]. It enhances Ca^{2+} and C-2-deoxyglucose uptake [11] and stimulates RNA synthesis in erythroid progenitors in vitro [12].

In this study we observed marked increases of PB BFU-E levels under rhEPO treatment in patients with anemia due to end-stage renal disease. Since rhEPO has also been shown to increase the concentration of erythroid progenitors in BM and to trigger BFU-E into S-phase [13], we conclude that the observed expansion of the BFU-E compartment in the circulation is not the effect of redistribution but reflects a direct or at least indirect stimulation of the precursors in the marrow pool.

Iron-depleted patients were found to have a hampered response to rhEPO therapy [6]. Our experience showed that a decrease in serum ferritin concentration was accompanied by a fall of PB BFU-E levels, suggesting that not only the mature cells of the erythroid lineage but also their immature progenitors are dependent on iron. A decrease of the erythroid progenitor cell pool may also explain a transient nonresponsiveness of patients to rhEPO therapy during bacterial infection [14], since we could demonstrate a severe reduction of BFU-E levels during pneumonia in 1 of our patients.

Most surprisingly in this study, we found that the stimulatory effect of rhEPO was not restricted to BFU-E but also included CFU-GM and CFU-MIX. In agreement with our findings, pharmacological doses of rhEPO trigger CFU-GM into S-phase and increase their concentration in BM of patients with end-stage renal failure [13]. In contrast to the marked impact on myeloid progenitors, the peripheral white blood cell count and the differential count remained unchanged under rhEPO therapy. This might be due to the fact that myelomonocytopoiesis is under a negative feedback control of differentiated cells [15] and particular cytokines, mainly the colony-stimulating factors, are required to form mature granulocytes and monocytes from their respective precursors.

The multilineage action of rhEPO in vivo is unexpected since rhEPO alone has been shown to stimulate only erythroid progenitor cells in vitro so far [5]. Since rhEPO administered to patients may act within a complicated network of a variety of humoral regulator molecules, we looked at the in vitro effects of rhEPO in combination with other cytokines. Among these experiments the most remarkable observation was a stimulation of myeloid progenitor cells by rhEPO if BM cells were pretreated with rhIL-3 for 3 h. At the moment we have no explanation for this finding nor can we say if this effect is a direct one or an indirect effect mediated through an action of accessory cells, but there may be some connection between this in vitro finding and the multilineage action of rhEPO in vivo.

In conclusion, our data on PB progenitor cells under rhEPO treatment and data from others on BM progenitors suggest that rhEPO is a multipotential hemopoietic growth factor that increases not only levels of BFU-E but also of CFU-GM and CFU-MIX. We also found some evidence for a multipotential effect of rhEPO in vitro but the underlying mechanisms remain unclear. Further investigations are necessary to clarify these mechanisms as well as a potential therapeutic value of this observation in states of disturbed hemopoiesis.

References

1 Erslev AJ: Humoral regulation of red cell production. Blood 1953;8:349.
2 Zins B, Drüecke T, Zingraff J, Bererhi L, Kreis H, Naret C, Delons S, Castaigne JP, Peterlonge F, Casadevall N: Erythropoietin treatment in anaemic patients on haemodialysis. Lancet 1986;ii:1329.
3 Jacobson LO, Goldwasser E, Fried W, Plazak L: The role of the kidney in erythropoiesis. Nature 1957;179:633.
4 McLeod DL, Shreeve MM, Axelrad AA: Improved plasma culture system for production of erythrocytic colonies in vitro: Quantitative assay method for CFU-E. Blood 1974;44:519.
5 Sieff CA, Emerson SG, Mufson A, Gesner TG, Nathan DG: Dependence of highly enriched human bone marrow progenitors on hemopoietic growth factors and their response to recombinant erythropoietin. J Clin Invest 1986;77:74–81.
6 Eschbach JW, Egrie JC, Downing MR, Browne JK, Adamson JW: Correction of the anemia of end-stage renal disease with recombinant human erythropoietin. Results of a combined phase I and II clinical trial. New Engl J Med 1987;316:73–78.
7 Bommer J, Müller-Bühl E, Ritz E, Eifert J: Recombinant human erythropoietin in anemic patients on hemodialysis. Lancet 1987;i:392.
8 Jacobs K, Shoemaker A, Rudersdorf R, Neill SD, Kaufmann RJ, Mufson A, Sheera J, Jones SS, Hewick R, Fritsch EF, Kawakita M, Shimizu T, Miyake T: Isolation and characterization of genomic and cDNA clones of human erythropoietin. Nature (Lond) 1985;313:806–810.

9 Fauser AA, Messner HA: Granuloerythropoietic colonies in human bone marrow, peripheral blood and cord blood. Blood 1978;52:1243–1248.
10 Dessypris EN, Krantz SB: Effect of pure erythropoietin on DNA synthesis by human marrow day 15 erythroid burst forming unit in short-term liquid culture. Br J Haematol 1984;56:295–306.
11 Sawyer S, Krantz S: Erythropoietin stimulates $^{45}Ca^{2+}$ uptake in Friend virus-infected erythroid cells. J Biol Chem 1984;259:2769–2774.
12 Bondurant M, Lind R, Koury M, Ferguson M: Control of globin gene transcription by erythropoietin in erythroblasts from Friend virus-infected mice. Mol Cell Biol 1985;5:675–683.
13 Dessypris EN, Grabner SE, Krantz SB, Stone MW: Effects of recombinant erythropoietin on the concentration and cycling status of human marrow hemopoietic progenitor cells in vivo. Blood 1988;72:2060–2062.
14 Kühn K, Nonnast-Daniel B, Grützmacher B, Grüner J, Pfäffl W, Baldamus CA, Scigalla P: Analysis of initial resistance of erythropoiesis to treatment with recombinant human erythropoietin; in Koch KM, Kühn K, Nonnast-Daniel B, Scigalla P (eds): Treatment of renal anemia with recombinant human erythropoietin. Contrib Nephrol. Basel, Karger, 1988, vol 66, pp 94–103.
15 Cline MJ, Golde DW: Cellular interactions in haematopoiesis. Nature 1979;277: 177–181.

K. Geissler, MD, 1st Department of Medicine, University of Vienna,
Lazarettgasse 14, A-1090 Vienna (Austria)

Recombinant Human Erythropoietin Stimulates Synthesis of Fetal Haemoglobin in Haemodialysed Patients with Anaemia due to End-Stage Kidney

Zygmunt Hanicki[a], Władysław Sułowicz[a], Marek Stepniewski[b], Marek Kuźniewski[a], Andrzej Kraśniak[a], Jerzy Kopeć[a], Urszula Cieszkowska[b], Katarzyna Stolarska[b]

Departments of [a]Nephrology and [b]Industrial Toxicology, Nicolaus Copernicus University School of Medicine, Kraków, Poland

The relevant literature, albeit abundant, lacks data concerned with the effect of recombinant human erythropoietin (rhuEPO) on the level of fetal haemoglobin (HbF) in the human. Thus, the aim of this study was to investigate whether and to what extent rhuEPO influences HbF concentration in subjects with terminal renal failure treated by repeated haemodialysis.

Material and Methods

The study included 10 patients (5 male, 5 female), aged 27–55 years, treated by chronic haemodialysis for 6–122 months (mean 58) due to end-stage kidney caused by chronic glomerulonephritis (9 patients) and by chronic bacterial tubulointerstitial nephritis (1 patient). The haemodialysis was carried out 3 times/week for 4–5 h. During the year preceding the study, all patients had blood transfusions (mean 4,295 ± 2,309 ml blood/patient) due to anaemia. One month before the onset of EPO administration, the patients received no blood transfusions.

HbF concentration was measured according to the method of Pembrey et al. [1]. The mean HbF level in 266 healthy persons assessed in this way was 0.350 ± 0.160%. The other parameters measured in the examined patients included total haemoglobin (Hb), haematocrit (Ht), and reticulocyte count. All these assessments were carried out prior to rhuEPO administration and on days 18, 36, 54, 72, 90 and 224 of the treatment.

rhuEPO ('Eprex', Cilag) was injected to the venous part of the line in a direct vicinity of the needle immediately after termination of the haemodialysis. The initial dose was 50 IU/kg b.w. 3 times/week. After increase in the Ht value to approximately 33%, the frequency of injections was reduced to twice a week. In 1 patient, who had not reached the

Table 1. Effect of rhuEPO on HbF concentration

	Before treatment 0	Days of treatment					
		18 A	36 B	54 C	72 D	90 E	224 F
x̄	0.354	0.390	0.573	0.697	0.578	0.294	0.330
SD	0.109	0.155	0.210	0.156	0.192	0.113	0.140

Differences statistically significant:
0:B $p < 0.02$ A:C $p < 0.02$ C:E $p < 0.001$ D:F $p < 0.01$
0:C $p < 0.001$ B:E $p < 0.01$ C:F $p < 0.001$
0:D $p < 0.02$ B:F $p < 0.01$ D:E $p < 0.01$

Table 2. Hb level, Ht and reticulocyte count during treatment with rhuEPO

		Before treatment	Days of treatment					
			18	36	54	72	90	224
Hb, g/dl	x̄	7.1	8.2	9.6	10.3	11.8	11.0	11.1
	SD	0.8	1.1	1.1	1.1	1.3	1.0	1.0
Ht, %	x̄	21.3	23.2	28.4	30.4	33.9	33.0	31.5
	SD	2.1	3.3	2.7	3.7	3.6	2.4	2.0
Reticulocytes, ‰	x̄	19.4	38.3	38.9	25.8	20.7	22.1	22.6
	SD	9.1	10.6	9.4	9.5	4.5	7.9	6.8

above value of Ht, the dose of rhuEPO was elevated to 75 IU/kg b.w. until the expected response was noted. This particular patient required intense iron supplementation. The obtained results were analysed statistically using Student's t test.

Results

Following administration of 50 IU rhuEPO/kg b.w. 3 times/week, the mean concentration of HbF in the studied patients gradually rose (table 1) and after 36, 54 and 72 days of treatment was significantly higher than prior to it ($p < 0.02$, $p < 0.001$, $p < 0.02$, respectively). The increase in HbF level was accompanied by a rise of total Hb, Ht and reticulocyte count (table 2).

When the rhuEPO dose was reduced to injections twice a week, HbF concentration systematically fell, to reach after 90 and 224 days of the entire treatment the level significantly lower than that found after 36 days ($p < 0.01$), 54 days ($p < 0.01$) and 72 days ($p < 0.01$), and close to that observed before rhuEPO administration. On the other hand, elevated values of total Hb and Ht achieved in the initial phase of rhuEPO treatment were maintained also after reduction of its dose. The reticulocyte count rapidly rose in the first weeks of the treatment, and its values assessed after 18 and 36 days of treatment were significantly higher than those before rhuEPO administration ($p < 0.001$).

Discussion

This study demonstrated a stimulatory effect of rhuEPO on HbF synthesis in haemodialysed patients with anaemia. In healthy individuals, secretion of EPO is continuous, resulting in a permanent activation of the erythroid progenitor cells. On the contrary, patients with end-stage kidney in which EPO production is usually residual, are supplemented with rhuEPO administered in a form of intermittent, relatively high pulses. This leads to a rapid, intense stimulation of the erythroid system, including F-cell precursors. Their activation results in increased synthesis of HbF. In contrast, physiological secretion of endogenous EPO or lower doses of rhuEPO retain the activity of F cells at the level observed in healthy persons.

Different HbF levels resulting from endogenous EPO secretion, as compared to treatment with rhuEPO, were reported in baboons. This led to speculations concerned with the induction of F-cell formation in patients with sickle cell anaemia. It has been shown that by stimulating the erythroid system, rhuEPO also causes an increase in HbF synthesis [2–5], similar to that induced by 5-azacytidine and α-amino-N-butyric acid, which revealed a beneficial effect on clinical manifestations of sickle cell anaemia and β-thalassaemia [6, 7].

Initial doses of rhuEPO used in this study brought about a gradual increase in both HbA and HbF levels, with the latter reaching the values higher than those in healthy subjects. The kinetics of the observed effect justifies a supposition that further supplementation with high pulses of rhuEPO would even more elevate the level of HbF. Such treatment was therefore avoided because of the expected known complications in this category of patients resulting from too high Ht.

Conclusions

To conclude: (1) in haemodialysed patients, HbF concentration does not change in comparison to healthy persons; (2) in the course of rhuEPO treatment, HbF level in haemodialysed patients increases, indicating activation of HbF synthesis and/or proliferation of F-cell precursors, and (3) administration of adjusted pulses of rhuEPO is justified in the treatment of sickle cell anaemia and β-thalassaemia.

References

1 Pembrey M, McWade P, Weatherall D: Reliable routine estimation of small amounts of foetal haemoglobin by alkaline denaturation. J Clin Pathol 1972;25: 738–740.
2 Stomatoyannopoulos G, Yeith R, Galanello R: HbF production in stressed erythropoiesis: observations and kinetic models. Ann NY Acad Sci 1985;445:188–197.
3 Al-Khatti A, Veith R, Papayannopoulou T, Fritsch E, Goldwasser E: Stimulation of fetal hemoglobin synthesis by erythropoietin in baboons. New Engl J Med 1987;317: 415–420.
4 Al-Khatti A, Papayannopoulou T, Veith R, Goldwasser E, Fritsch E: Fetal hemoglobin and erythropoietin. New Engl J Med 1988;318:450.
5 Barber H: Fetal hemoglobin and erythropoietin. New Engl J Med 1988;318:449.
6 Kolata G: Fetal hemoglobin genes turned on in adults. Science 1982;218:1295–1296.
7 Constantoulakis P, Papayannopoulou T, Stomatoyannopoulos G: Alpha-amino-N-butyric acid stimulates fetal hemoglobin in the adults. Blood 1988;72:1961–1967.

Prof. Dr. Zygmunt Hanicki, Department of Nephrology, Nicolaus Copernicus University School of Medicine, Kopernika 15, PL–31-501 Kraków (Poland)

Erythropoietin Production in Chronic Renal Disease before and after Transplantation

Armin Kurtz, Kai-Uwe Eckardt

Physiologisches Institut der Universität, Zürich, Switzerland

A moderate to severe anemia is virtually a hallmark of chronic renal failure. This anemia develops parallel to the reduction in excretory renal function and is more or less independent of the underlying renal disease. However, there are exceptions from this rule and in particular patients with polycystic kidneys generally remain far less severely anemic despite their loss of excretory renal function. The etiology of renal anemia is considered to be multifactorial. A variety of causal factors, including chronic blood losses, a shortened red cell survival time, putative uremic inhibitors of erythropoiesis and myelofibrosis due to hyperparathyroidism have to be considered. The main reason for the anemia during end-stage renal disease (ESRD), however, appears to be an inappropriately low production rate of erythropoietin (EPO), the main humoral regulator of red cell formation. The predominant role of a relative EPO deficiency in the pathogenesis of renal anemia has become most obvious from the recent clinical experience that treatment with recombinant EPO is able to overcome this anemia in almost all patients with ESRD [1, 2].

The only alternative to effectively treat renal anemia is kidney transplantation, after which the restoration of excretory function is followed by an improvement of erythropoiesis. Some patients not only correct their anemia, but even develop a polycythemia after renal transplantation, which raises the risk of thromboembolic complications.

In order to assess the mechanisms and the role of a disturbed EPO formation in these hematopoietic disorders associated with renal disease, it seems reasonable to analyze EPO production before and after renal transplantation and to compare it with the situation in healthy beings.

Regulation of EPO Production in Healthy Beings

In humans like in a number of laboratory animals two production sites for EPO have been recognized, namely the liver and the kidneys [3]. While the liver is considered as the source of EPO during fetal and perhaps neonatal life, the kidneys elaborate about 90% of the EPO in children and adults. In order to characterize the regulation of EPO levels it seems reasonable therefore to confine oneself to *renal* EPO formation. Renal EPO formation is regulated by a variety of factors. Presumably mediated through the effect of other *hormones*, EPO production is coupled to the energetic situation of the organism. Lowered energy consumption, e.g. during starvation or after hypophysectomy, is accompanied by a reduction in EPO production rates. Conversely, an enhancement of energy turnover after application of thyroid hormone, growth hormone or androgens is accompanied by increased EPO production rates [4]. Under normal endocrine conditions, *oxygen supply* to the tissues is the dominant control factor for EPO production. Since oxygen delivery is determined by the arterial oxygen content and by the oxygen affinity of hemoglobin, both parameters affect EPO production: EPO production is inversely related to the arterial oxygen content and positively related to oxygen affinity [5]. Taken a normal oxygen dissociation, the arterial oxygen tension determining *oxygen saturation* [6] and the *hemoglobin concentration* determining the oxygen carrying capacity [7, 8] are thus the main determinants of EPO production.

There is no clear evidence that the half-life time of EPO, in humans in the range between 6 and 9 h, is subject to any regulation. Furthermore, no stores for EPO exist in the kidneys. Therefore any change in circulating EPO levels appears to directly reflect an alteration in EPO production rate.

The adaptation of EPO production to the oxygen delivery obviously requires perception of the oxygen delivery, comparison with a demand value and transduction of the signal into appropriate EPO production. It is reasonable to assume that the sensing of oxygen delivery occurs in the kidneys itself, because even isolated perfused kidneys produce EPO in an inverse relationship to the oxygen delivery [9]. Probably the local oxygen tension in certain areas of the kidney cortex represents the signal that determines EPO formation. Actually, the oxygen tension in the kidney appears to be very suitable to register oxygen delivery representative for the whole organism, because it is relatively insensitive to local changes in blood flow. This is because a reduction in renal perfusion does not only reduce oxygen supply but at the same time via a reduction in filtration rate also lowers oxygen consumption. In fact it has been shown that a reduction in renal blood flow is only a minor stimulus for EPO formation when compared to a reduction in hemoglobin concentrations [10].

Strong evidence suggests that EPO is produced by peritubular cells, which have, however, not yet been identified unequivocally [11, 12]. Possible candidates are capillary endothelial or interstitial mesenchymal cells. It is not yet known whether these EPO-producing cells themselves function as 'oxygen sensors'. In any case it appears as if normal oxygen sensing and its transduction into appropriate EPO production requires intact tubular, in particular proximal tubular function [13]. It remains to be clarified whether upon insufficient oxygen supply the tubules generate signals which are transferred to the EPO-producing cells in their neighborhood, or whether the tubules merely amplify a local reduction of oxygen tension in the renal interstitium by their high rate of oxygen consumption.

EPO Production during ESRD

The alteration of EPO production during ESRD is most impressively reflected when plasma EPO levels are plotted against hemoglobin concentrations (fig. 1). In contrast to patients with nonrenal anemias, reductions in hemoglobin values during ESRD are not accompanied by an increase of EPO levels. EPO levels generally remain within, slightly above or even below the normal range of nonanemic individuals. Conversely to the inverse relationship between EPO levels and hemoglobin in anemias not associated with kidney failure, there appears to be even a positive relationship between EPO and hemoglobin in ESRD, suggesting that hemoglobin concentrations are determined by a given amount of EPO produced instead of circulating EPO levels responding to a fall of red cell mass.

Principally, there are two possible explanations for this inappropriately low EPO formation. First it is conceivable that chronic structural damage to the kidneys destroys the cells normally producing EPO and the low serum EPO levels would then simply reflect a loss of production capacity. Alternatively, it is also possible that the capability to produce EPO is preserved, but that those mechanisms which normally adjust EPO production to oxygen supply are disturbed in ESRD. Presently we cannot clearly distinguish between these two possibilities and their relative importance may vary according to the etiology and the progression of renal disease. At least there are a number of indications, suggesting that a defective 'oxygen sensor' rather

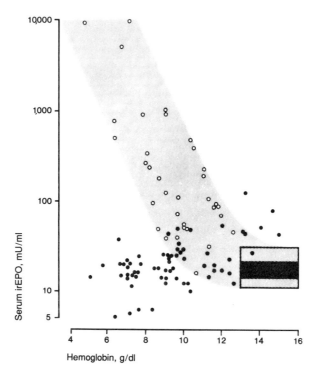

Fig. 1. Relationship between serum immunoreactive (ir) EPO levels and hemoglobin concentrations in hypo- and hyperregenerative nonrenal anemias (○) and in patients with chronic renal failure (●) (excluding patients with polycystic kidney disease). The rectangle depicts interquartile range (dark stippled) and 95% confidence range of EPO levels in nonanemic healthy adults. Values in renal failure patients were determined prior to kidney transplantation [adapted from 21 and 22].

than a destruction of EPO producers is primarily responsible for the relative EPO deficiency that accounts for renal anemia. For instance, an acute reduction in arterial oxygen tensions in patients with chronic renal failure may cause a considerable rise in serum EPO. This has been observed after exposing uremic patients to high altitude [14] and also in the course of cardiopulmonary complications of renal failure, e.g. pulmonary edema [15] (fig. 2). Furthermore, an acute blood loss superimposed on the preexisting anemia was also found to induce some increase in serum EPO in ESRD patients [16]. Together these observations could be indicative for a sufficient EPO-producing capacity of diseased kidneys, which is, however, not utilized due to a marked desensitization of the renal 'oxygen sensor'.

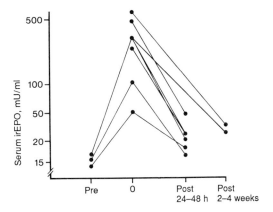

Fig. 2. Immunoreactive (ir) EPO levels in children with renal anemias before, during and after acute hypoxic episodes [adapted with permission from 15]. Although EPO levels are inappropriately low for the given hematocrit, additional reduction in oxygen supply results in a marked increase in EPO.

Another specific example where the capability of diseased kidneys to produce EPO has been clearly demonstrated is *polycystic kidney disease* [17, 18]. Arteriovenous concentration differences for EPO exist in these cystic kidneys (fig. 3) and peripheral serum EPO values are on average twofold higher than in other patients with renal disease. However, EPO production by the cystic kidneys appears not to be regulated in inverse relation to circulating hemoglobin concentrations. If at all, hemoglobin and serum EPO are positively correlated also in these patients [18], indicating that circulating EPO levels do not properly respond to a fall in red cell mass. Thus again it appears that a disturbance in 'oxygen sensing' accompanies the cystic deformation of these kidneys, whereas the capability to synthesize EPO is retained.

EPO Production after Renal Allotransplantation

Clinical experience teaches that renal anemia is corrected within several weeks after successful renal allotransplantation. If we accept that it is mainly an inappropriate EPO production that leads to anemia in ESRD, then consideration of EPO production after renal transplantation gains interest. In fact, provided immediate excretory graft function, serum EPO levels rise

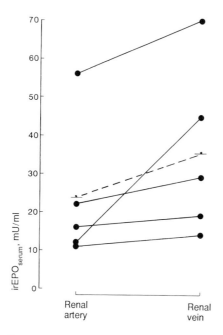

Fig. 3. Arteriovenous concentration differences for EPO in polycystic kidneys [reproduced with permission from 17].

within 3–4 days after transplantation [19–21]. When hemoglobin concentrations subsequently have normalized, serum EPO declines to normal values [22, 23]. These findings suggest that the transplant is capable to register the anemia of ESRD and to respond with an increase in EPO production. The oxygen dependency of EPO formation by functioning transplanted kidneys becomes even more obvious when patients undergo acute blood loss in the posttransplant period. As illustrated in figure 4, left panel, a temporal reduction in hemoglobin results in a further increase in EPO formation, which is quantitatively in accordance with the physiological exponential inverse relationship between EPO and hemoglobin concentrations.

Regarding the mechanisms responsible for a disturbance of EPO production during kidney failure, the analysis of EPO formation after transplantation is of particular interest in those cases where the graft initially lacks excretory function. In this situation, EPO formation was found to be rather unpredictable and variable [21]. In some cases temporal increases occur and EPO levels may display an irregular 'spiking' pattern with no relationship to

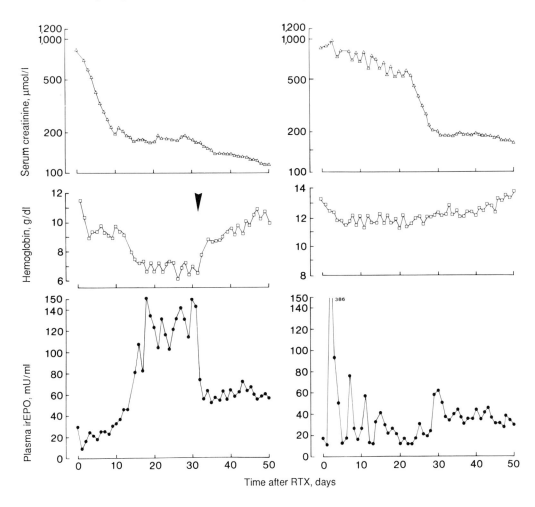

Fig. 4. Time course of serum creatinine (upper), hemoglobin (middle) and serum EPO (lower) concentrations in 2 patients following renal allotransplantation [reproduced with permission from 21]. *Left:* A patient is shown with immediate graft function, in whom EPO increases parallel to the onset of excretory function. A temporary reduction in hemoglobin due to gastrointestinal bleeding is accompanied by a further increase in EPO (arrow indicates transfusion of two units of red cells). *Right:* A patient is shown with delayed graft function, in whom EPO displays marked irregular and inconsistent increases during the period of excretory graft failure. Note that a continuous increase in EPO occurs also in this patient following the onset of excretory graft function.

hemoglobin concentrations (fig. 4, right panel). In others no evidence for EPO production by transplanted kidney is obtained at all and also a reduction in hemoglobin concentrations does not evoke an increase in EPO. As soon as excretory function commences, EPO production increases in primary inactive kidneys or stabilizes in those cases where inconsistent elevations occurred during graft failure. In both instances, EPO levels after onset of excretory function reach the values typical for cases with immediate graft function. Irrespective of the different behavior of EPO formation in individual cases, it appears again as if during kidney failure the production capacity for EPO may be preserved but becomes decoupled from oxygen-dependent regulation. Furthermore, the temporal concurrence between the onset of excretory renal function and a regular oxygen dependency of EPO formation in all cases may indicate that ongoing excretory function in the kidney is a prerequisite for the regulation of EPO. Based on what was previously said about the interrelationship between tubular cells and EPO-producing cells in their vicinity, it is conceivable that a lack of tubular oxygen consumption or biochemical signals derived from the tubules impairs an appropriate increase in EPO in response to hypoxia.

EPO Production during Posttransplant Polycythemia

In about 10% of the patients undergoing successful renal allotransplantation a polycythemia develops, which in turn may give rise to thromboembolic complications. Serum EPO concentrations in these patients were reported to be elevated [24, 25] and in some cases serial determinations showed a sustained elevation of EPO after transplantation despite an increase in hemoglobin [23]. The question arises about the source of EPO in these patients. Both the native kidneys, which are normally left in situ, and the graft have to be considered. In some instances an association with rejection, artery stenosis or hydronephrosis of the transplanted kidney has been reported [cf. 24]. On the other hand, posttransplant erythrocytosis is generally only observed in patients who have not been nephrectomized prior to transplantation. Furthermore, it is known from clinical experience that exstirpation of the native kidneys most often cures posttransplant polycythemia [26], suggesting a substantial role of the native kidneys as a source for inappropriate (high) EPO secretion. This clinical inference is supported by multiple site estimates of EPO in posttransplant patients with polycythemia, demonstrating significant arteriovenous concentration differences for EPO

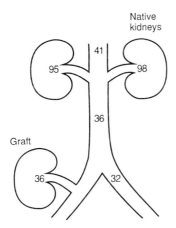

Fig. 5. Multiple site estimates of immunoreactive EPO levels in samples obtained by selective venous catheterization of a patient developing polycythemia after renal transplantation. Marked elevations of EPO are found in the veins draining the native kidneys [courtesy of Prof. Lesky, Geneva].

in the native kidneys [24, 25] (fig. 5). At first view it appears somewhat contradicting that after transplantation EPO formation by diseased kidneys appears to induce polycythemia, whereas before transplantation inappropriately low EPO formation is generally considered the main reason for the development of renal anemia. However, EPO determinations in nonrenal patients with secondary erythrocytosis clearly indicate that only a slight increase in EPO is sufficient to maintain erythrocytosis [27]. Thus it is possible that the amount of EPO produced by the diseased kidneys is insufficient to overcome the anemia, but once the anemia is corrected by additional EPO formation of the transplant, it may suffice to maintain an increased rate of erythropoiesis. In addition to what was mentioned about the failure of diseased kidneys to sufficiently enhance their EPO formation under hypoxia, these observations indicate that also the ability to down-regulate EPO production under conditions of increased red cell mass is disturbed. Again, this finding supports the concept that kidneys without excretory function are capable to produce EPO, but that a defect of the renal 'oxygen sensor' is a major cause for inappropriate EPO formation during renal disease.

References

1 Eschbach JW: The anemia of chronic renal failure: Pathophysiology and the effects of recombinant erythropoietin (Nephrology Forum). Kidney Int 1989;35: 134–148.
2 Cotes PM, Pippard JM, Reid CDL, Winearls CG, Oliver DO, Royston JP: Characterisation of the anemia of chronic renal failure and the mode of its correction by a preparation of erythropoietin (rhuEPO). An investigation of the pharmacokinetics of intravenous rhuEPO and its effect on erythrokinetics. Q J Med 1989;70: 113–137.
3 Kurtz A, Eckardt K-U, Tannahill L, Bauer C: Regulation of erythropoietin production. Contrib Nephrol. Basel, Karger, 1988, vol 66, pp 1–16.
4 Jelkmann W: Renal erythropoietin: properties and production. Rev Physiol Biochem Pharmacol 1986;104:139–215.
5 Lechermann B, Jelkmann W: Erythropoietin production in normoxic and hypoxic rats with increased blood O_2 affinity. Respir Physiol 1985;60:1–8.
6 Eckardt K-U, Boutellier U, Kurtz A, Schopen M, Koller EA, Bauer C: Rate of erythropoietin formation in humans in response to acute hypobaric hypoxia. J Appl Physiol 1989;66:1785–1788.
7 Cotes PM: Immunoreactive erythropoietin in serum. Evidence for the validity of the assay method and the physiological relevance of estimates. Br J Haematol 1982;50: 427–438.
8 Erslev AJ, Wilson J, Caro J: Erythropoietin titers in anemic, nonuremic patients. J Lab Clin Med 1987;109:429–433.
9 Scholz H, Schurek HJ, Eckardt K-U, Kurtz A, Bauer C: Oxygen-dependent erythropoietin production by the isolated perfused rat kidney (submitted).
10 Pagel H, Jelkmann W, Weiss C: A comparison of the effects of renal artery constriction and anemia on the production of erythropoietin. Pflügers Arch 1988; 413:62–66.
11 Lacombe C, Da Silva JL, Bruneval P, Fournier JG, Wendling F, Cassadevall N, Camillieri JP, Bariety J, Varet B, Tambourin P: Peritubular cells are the site of erythropoietin synthesis in the murine hypoxic kidney. J Clin Invest 1988;81:620–623.
12 Koury ST, Bondurant MC, Koury MJ: Localization of erythropoietin-synthesizing cells in murine kidneys by in situ hybridization. Blood 1988;71:524–527.
13 Eckardt K-U, Kurtz A, Bauer C: Regulation of erythropoietin formation is related to proximal tubular function. Am J Physiol 1989;256:F942–F947.
14 Blumberg A, Keller H, Marti HR: Effect of altitude on erythropoiesis and oxygen affinity in anaemic patients on maintenance dialysis. Eur J Clin Invest 1973;3:93–97.
15 Chandra M, Clemons GK, McVicar MI: Relation of serum erythropoietin levels to renal excretory function: evidence for lowered set point for erythropoietin production in chronic renal failure. J Pediatr 1988;113:1015–1021.
16 Walle AJ, Wong GY, Clemons GK, Garcia JF, Niedermayer W: Erythropoietin-hematocrit feedback circuit in the anemia of end-stage renal disease. Kidney Int 1987;31:1205–1209.
17 Eckardt K-U, Möllmann M, Neumann R, Brunkhorst R, Burger H-U, Lonnemann G, Scholz H, Keusch G, Buchholz B, Frei U, Bauer C, Kurtz A: Erythropoietin in polycystic kidneys. J Clin Invest 1989;84:1160–1166.

18 Chandra M, Miller ME, Garcia JF, Mossey RT, McVicar M: Serum immunoreactive erythropoietin levels in patients with polycystic kidney disease as compared with other hemodialysis patients. Nephron 1985;39:26–29.
19 Besarab A, Caro J, Jarell BE, Francos G, Erslev AJ: Dynamics of erythropoiesis following renal transplantation. Kidney Int 1987;32:526–536.
20 Rejmann ASM, Grimes AJ, Cotes PM, Mansell A, Joekes AM: Correction of anaemia following renal transplantation: Serial changes in serum immunoreactive erythropoietin, absolute reticulocyte count and red-cell creatine levels. Br J Haematol 1985;61:421–431.
21 Eckardt K-U, Frei U, Kliem V, Bauer C, Koch KM, Kurtz A: Role of excretory graft function for erythropoietin formation after renal transplantation. Eur J Clin Invest 1990;20:564–574.
22 Keusch G, Kurtz A, Fehr J, Eckardt K-U, Frei D, Bauer C, Binswanger U: Erythropoiese und Serumerythropoietinkonzentration vor und nach Nierenallotransplantation. Nephron 51(suppl 1):29–33.
23 Sun CH, Ward HJ, Wellington LP, Koyle MA, Yanagawa N, Lee DBN: Serum erythropoietin levels after renal transplantation. New Engl J Med 1989;321:151–157.
24 Dagher FJ, Ramos E, Erslev AJ, Alongi SV, Karmi SA, Caro J: Are there native kidneys responsible for erythrocytosis in renal allorecipients? Transplantation 1979;28:496–498.
25 Thevenod F, Radtke HW, Grützmacher P, Vincent E, Koch KM, Fassbinder W: Deficient feedback regulation of erythropoiesis in kidney transplant patients with polycythemia. Kidney Int 1983;24:227–232.
26 Janhez LE, Da Fonseca JA, Chocair PA, Maspes V, Sabbaga E: Polycythemia after kidney transplantation. Urol Int 1977;32:382–392.
27 Cotes PM, Dore JC, Yin JAL, Lewis SM, Messinezy M, Pearson TC, Reid C: Determination of serum erythropoietin in the investigation of erythrocytosis. New Engl J Med 1986;315:283–287.

Armin Kurtz, MD, Physiologisches Institut der Universität Zürich,
Winterthurerstrasse 190, CH–8057 Zürich (Switzerland)

The Role of Erythropoietin in Patients with Anemia and Normal Renal Function

W. Oster[a], R. Mertelsmann[b,1]

[a]Behringwerke AG, Marburg, and [b]Department of Internal Medicine I, University of Freiburg i.Br., FRG

Biochemistry and Molecular Biology

Erythropoietin (EPO) is an acidic sialylglycoprotein with an isoelectric point close to 4.5, consisting of a 166 amino acid polypeptide with a molecular mass of the protein backbone of 18,398 daltons. Native EPO is heavily glycosylated, exhibiting a complex polyantennary sugar structure, so that the final secreted product has a molecular mass ranging from 34,000 to 39,000 daltons. The carbohydrate structure of EPO has been shown to be important for its biological activity in vivo. Enzymatic removal of either the terminal sialic acid moieties or the N-linked oligosaccharide chains results in an almost total loss of in vivo activity with no loss of in vitro biological activity, which might be explained by a possible prevention of rapid hepatic clearance by the carbohydrate structure [10]. EPO contains 3 asparaginase-linked carbohydrate chains and, in addition, possibly several O-linked glycosylation sites [7].

The entire coding region of the single copy EPO gene is contained in a 5.4-kb HindIII-BamHI fragment. The gene contains 4 intervening sequences and at least 5 exons [13, 19]. The human EPO gene has been assigned to chromosome 7q11 [16]. A strongly expressed mRNA species has been identified in human fetal liver corresponding to 1,600 nucleotides in length.

---

[1] We would like to express our gratitude to Ms. G. Seitz for the excellent preparation of the manuscript.

Low-level mRNA of identical size was detected in adult liver, and transcripts of 2,000 nucleotides were weakly expressed in both fetal and adult liver.

Biological Effects and Sources

EPO is the principal hormone involved in the regulation and maintenance of a physiological level of circulating erythrocyte mass in man [25]. Although biogenesis of the hormone is still not clearly defined, EPO is known to be produced in substantial amounts by capillary endothelial cells and a subset of interstitial cells of the adult kidney [15, 27] and by the liver during fetal life [31]. Under normal conditions, 10–15% of all synthesis of EPO occurs in the liver of the adult [2]. In man, EPO is maintained in the circulation at a concentration of about 15–30 units/l of serum (EPO activity is expressed in terms of activity of an International Reference Preparation made available by the National Institute for Medical Research, London) or about 0.01 nM under normal physiological conditions, and it is excreted in the urine at very low concentrations (about 2 units daily) [3].

EPO exerts its biological effects by attachment to specific receptors on target cells. Cross-linking studies with radio-iodinated EPO initially show that the receptor is comprised of 2 polypeptides with molecular masses of 85,000 and 100,000 daltons. However, digestion experiments suggested that both polypeptides may be products of the same or very similar genes [26]. Recently the murine EPO receptor has been cloned and showed a single affinity with a dissociation constant of approximately 240 pM [4]. The EPO receptor cDNA expressed in COS cells generated both a high-affinity (30 pM) and a low-affinity (210 pM) receptor.

There are conflicting data on the binding site of EPO. Recently it was shown that an interior peptide structure (1199-129) may be involved in the binding of the molecule, whereas other results favored a role for the carboxylterminal region of the molecule in the attachment to the receptor [9].

Clinical Experience

EPO in Patients with Renal Deficiency

Lack of EPO production has been shown to be the major cause of anemia in patients with end-stage renal disease and progressive renal failure [25]. Several clinical trials provide evidence for high efficacy of recombinant

Fig. 1. Treatment scheme for EPO administration. EPO was given as an intravenous bolus injection at the escalating doses indicated, twice weekly for the total duration of 14 weeks.

human EPO (rhEPO) in the treatment of anemias associated with deficient endogenous EPO supply [8, 18, 30]. EPO treatment has proved to be beneficial for these patients in several aspects, including improvement of quality of life, prevention of transfusion-related risks of infection and iron overload among other benefits.

EPO in Patients with Intact Renal Function

The effects of EPO on erythropoiesis in anemic disease states unrelated to renal insufficiency have, until recently, been controversially discussed. Recent data provide evidence for efficacy of EPO in treating anemia of rheumatoid arthritis (RA). Hematocrits in these patients were normalized with doses of rhEPO ranging from 150 to 200 U/kg given intravenously 3 times weekly [21].

A substantial proportion (ca. 30%) of AIDS patients become transfusion-dependent under Zidovudine (AZT) treatment. Preliminary reports show that patients with baseline endogenous EPO serum levels <500 mU/ml respond to rhEPO at doses up to 300 U/kg/week with decreased transfusion requirements. Studies with escalating doses of EPO starting at 400 U/kg/dose 3 times/week and increasing to a maximum at 1500 U/kg/dose showed dose-dependent efficacy of EPO, producing a response when lower doses failed. No side effects have been reported in these studies [17].

First experiences with EPO in the treatment of anemia of malignancy were gathered from patients with neoplastic bone marrow involvement by low-grade non-Hodgkin's lymphoma (lg NHL) and multiple myeloma (MM) [24]. Treatment with the study drug was preceded by a 2-week period of placebo administration. EPO was started at a comparatively low dose of 150 U/kg body weight (b.w.) twice weekly, given as an intravenous bolus injection. If the increase in hemoglobin (Hb) remained less than 1 g/dl, the EPO dose was escalated after 6 weeks up to 300 U/kg b.w. (fig. 1). Applying

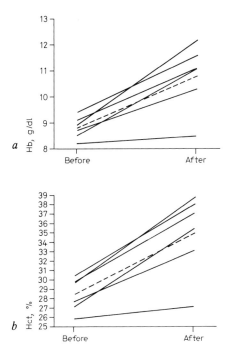

Fig. 2. a Increase of Hb values of 6 patients as measured before and after EPO therapy. Dashed line provides mean values of 6 patients. *b* The same information on Hct.

the same criteria, another escalation step followed after another 4 weeks utilizing an EPO dose of 450 U/kg b.w.

Four of 5 patients with lg NHL and initial Hb values under 10 g/dl showed a homogeneous response to EPO resulting in a mean elevation of Hb levels of 2.1 g/dl (range 0.8–3.3 g/dl) during therapy. One patient with severe extensive disease showed only a minor response with a Hb increase of 0.3 g/dl (fig. 2). All responding patients became free of the need of erythrocyte transfusions and amelioration of erythropoietic parameters was associated with a reduction of serum ferritin values by more than 50% in most cases, which were initially high due to polytransfusion. EPO serum levels of all patients were found to be elevated (range 74–202 mU/ml) prior to commencing the trial.

Two patients with chronic myelogenous leukemia (CML) and suffering from anemia (Hb < 10 g/dl) were studied for potential responsiveness to EPO.

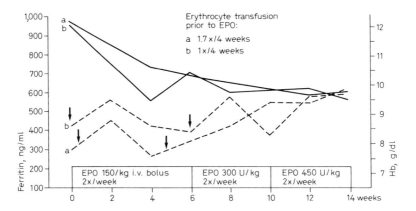

Fig. 3. Courses of Hb (---) and serum ferritin (——) in 2 patients (a and b) with CML under EPO treatment. Frequency of erythrocyte transfusions (500 ml washed erythrocytes) is given for each patient on a basis of 4 weeks. Arrows indicate erythrocyte transfusions under EPO treatment.

Using the same route of administration for EPO we found a significant reduction of ferritin in both cases and an increase in Hb and hematocrit (Hct) with a slope which was lower as compared to the patients with lg NHL (fig. 3). EPO serum levels in these 2 anemic patients were initially found to be normal.

Patients with anemia associated with multiple myeloma (MM) showed responses to EPO treatment very similar to those observed in patients with lg NHL. Most importantly, these patients developed increased platelet counts under EPO by more than 75% which adds significantly to the therapeutic success in this disease state, often complicated by thrombocytopenic bleeding. With the recent observation of EPO-induced expansion of megakaryocytic progenitor cells (CFU-Meg) [5] and the identification of EPO receptors on megakaryocytes, this unexpected phenomenon finds a possible explanation.

Most patients showed response to EPO in the late phase of the second dose level and more prominently on the highest dose level. Clinical experience with other hematopoietic growth factors in cancer patients showed that constant drug administration is most effective [12]. The pharmacokinetics were examined after intravenous and subcutaneous single dose of rhEPO in cynomolgous monkeys and man. The resulting pharmacokinetic data are shown (for cynomolgous monkeys) in figure 4. There was a 10-fold increase of serum half-life of EPO when given via subcutaneous administration as

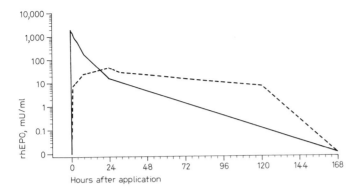

Fig. 4. Pharmacokinetics of rhEPO after single injection (70 U) in cynomolgous monkeys. Comparison of intravenous (——) and subcutaneous (– – –) administration.

compared to intravenous bolus injection. These results led us to change the treatment protocol to a subcutaneous injection at a dose of 10,000 U/day, 5 days/week.

For the indications lg NHL and MM, this new treatment modality was associated with a faster increase of Hb and Hct. Currently several malignant disease states associated with anemia are evaluated for the response to exogenous EPO administered as a subcutaneous injection. Besides the comparison to the intravenous route of administration, this study addresses the question of relationship between EPO dose and efficacy. The results of this study, which is ongoing in our institution, have to be awaited to draw reliable conclusions about optimum dose and route of administration of EPO.

We also studied the effects of EPO given as an i.v. bolus injection in patients with myelodysplastic syndromes (MDS) including refractory anemia with exessive blasts (RAEB) and refractory anemia with excessive blasts-transformed (RAEB-T). At the current state of investigation, 10 patients are evaluable for efficacy of treatment. There was only a minor or no increase in Hb, Hct and RBC blood values. However, 4 patients showed a reduction of their erythrocyte transfusion requirements (>50%). The effective stimulation of erythropoiesis by EPO was also reflected by the reduction of serum ferritin values in some of these patients. Two patients experienced a reduced need for platelet transfusions (>50%), which was maintained for several weeks after EPO had been discontinued. However, this clinical improvement was not associated with a significant rise of platelet counts. Overall response was less remarkable as compared to patients with Ig NHL. This discrepancy may relate to EPO serum levels which were found to be much higher (range ca. 400–1,200 mU/ml) in patients with MDS. This may indicate that the etiology of anemia in these patients is multifactorial and includes dyserythropoiesis which may be more difficult to overcome than a sole EPO deficiency.

First experiences with EPO given as a subcutaneous injection at dose and duration mentioned above include anemic patients with MDS. Some of these patients show responsive erythropoiesis achieving a stable plateau of Hb above 11 g/dl. However, the therapeutic potential of EPO in patients with MDS needs to be further investigated.

Preliminary experience of other investigators supports our findings [20]. Patients with anemias due to bone marrow infiltrating NHL and MM tend to show the most impressive results, whereas data on patients with MDS remain variable. Several investigators in Europe and North America are currently investigating the potential benefit of EPO for MDS. Possibly the stratification for the various subtypes of MDS (refractory anemia, RAEB, RAEB-T) may throw light on the heterogeneous response patterns reported.

Another concern of oncologists associated with anemias is to obtain means to avoid transfusion therapy when an anemic period can be foreseen, e.g. following chemotherapy. In one well-documented case we showed that EPO has the potential to correct anemia induced by cisplatinum-based treatment [23]. Moreover, we found that proliferation of erythropoiesis continued under EPO although further chemotherapy was administered and the following nadir counts of RBC and platelets were clearly higher as compared to prior nadir counts after identical chemotherapy without EPO (Hb 7.0 vs. 8.9 g/dl; platelets 9 vs. $59 \times 10^3/\mu l$). Again a more than 60% reduction of serum ferritin values was associated with the rise of Hb. EPO serum level was determined before starting the trial and was found to be elevated at 215 mU/ml which, for the severe degree of anemia, may be inadequate.

The potential of EPO to act on erythropoiesis while chemotherapy is continued has also recently been mentioned in a preliminary report by other investigators [11].

Patients with a variety of malignant disease states associated with anemia have been enrolled in EPO trials. Some case reports have been mentioned. Efficacy of EPO has been shown to be potentially beneficial in pure red cell anemia [28]. Other casuistics report failure of EPO to correct anemia in Fanconi syndrome [29] and Blackfan-Diamond syndrome [own unpublished results]. These data are not proof of efficacy or failure of EPO in the indications mentioned but may indicate a field of further investigations.

EPO is currently studied in patients after autologous and allogeneic bone marrow transplantation. A first preliminary report shows efficacy of EPO to induce a faster increase of reticulocyte counts and Hct in the phase of recovery [6].

In contrast to adverse events reported from patients with renal disease, cancer patients tolerate EPO treatment extremely well. No serious side effect has so far been reported from any of the ongoing clinical trials. This includes our experience from a cohort of more than 100 patients who received EPO or are currently under treatment with EPO.

The lack of hypertension and other side effects in these patients may relate to the shorter duration of anemic periods which do not cause severe anemic vasodilation. This mechanism, as well as increased vascular risks, may be the major reasons for complications in EPO-treated patients with renal disease [1].

The mechanism of action of EPO in anemias associated with malignancy is controversial. It has recently been reported that EPO production by these patients is considerably lower as compared to patients with a similar degree of anemia caused by iron deficiency [22]. This finding may be supported by EPO serum levels mentioned above. The degree of anemia observed does not seem to correlate to the EPO response in these patients. It remains unclear at the moment why EPO production in these patients is reduced or inhibited. Other investigations favor a suppression of erythropoiesis which may be involved in some cancer patients, e.g. by TNF-α. It has recently been shown that TNF-α-inhibited erythropoiesis in mice can be reversed by EPO treatment [14]. However, TNF serum levels observed in a series of cancer patients did not correlate with the degree of anemia in these individuals [24].

Physicians and scientists have been cooperating to characterize and isolate the hormone that controls erythropoiesis named EPO for more than a century [25]. With EPO available as a therapeutic means, the spectrum of indications where this hormone may be used for the benefit of the patient extends far beyond its application in anemia of renal disease.

References

1 Adamson JW, Eschbach JW: The use of recombinant human erythropoietin in man; in Mertelsmann R, Herrmann F (eds): Hematopoietic Growth Factors in Clinical Application. New York, Dekker (in press).

2   Chandra M, Miller ME, Garcia JF, Mossey RT, McVicar M: Serum immunoreactive erythropoietin levels in patients with polycystic kidney disease as compared with other hemodialysis patients. Nephron 1985;39:26–29.
3   Cotes PM: Immunoreactive erythropoietin in serum. Br J Haematol 1982;50:427–438.
4   D'Andrea AD, Lodish HF, Wong GG: Expression cloning of the murine erythropoietin receptor. Cell 1989;57:277–285.
5   Dessypris EN, Graber SE, Krantz SB, Stone WJ: Effects of recombinant erythropoietin on the concentration and cycling status of human hematopoietic progenitor cells in vivo. Blood 1988;72:2060–2065.
6   Ebell W, Bucsky P, Diedrich H, Seidel J, Stoll M, Freund M, Sens B, Tischler J, Brune T, Riehm H, Poliwoda H, Link H: Use of recombinant human erythropoietin after bone marrow transplantation. Mol Biother 1989;1:53A.
7   Egrie JC, Strickland TW, Lane J, Aoki K, Cohen AM, Smalling R, Trail G, Lin FK, Browne JK, Hines DK: Characterization and biological effects of recombinant human erythropoietin. Immunobiology 1986;172:213–224.
8   Eschbach JW, Kelly MR, Haley NR, Abels RI, Adamson JW: The treatment of the anemia of progressive renal failure with recombinant human erythropoietin. New Engl J Med 1989;321:158–163.
9   Fibi M, Stüber W, Hintz-Obertreis P, Krumwieh D, Siebold B, Zettlmeissel G, Küpper H: Evidence for the location of the receptor binding site of human erythropoietin of the carboxylterminal domain. Blood (in press).
10 Goldwasser E, Kung CKH, Eliason J: On the mechanism of erythropoietin-induced differentiation. J Biol Chem 1974;249:4202–4206.
11 Henry DH, Rudnick SA, Bryant E, Abels RI, Danna RP, Staddon AP, Mason BA: Preliminary report of two double-blind, placebo-controlled studies using human recombinant erythropoietin in the anemia associated with cancer. Blood 1989;74 (suppl 1):6A.
12 Herrmann F, Schulz G, Lindemann A, Meyenburg W, Oster W, Krumwieh D, Mertelsmann R: Hematopoietic responses in patients with advanced malignancy treated with recombinant human granulocyte-macrophage colony-stimulating factor. J Clin Oncol 1989;7:159–167.
13 Jacobs K, Shoemaker C, Rudersdorf R, Neill SD, Kaufman RJ, Mufson A, Seehra J, Jones SS, Hewick R, Fritsch EF, Kawahita M, Shimizu T, Miyake T: Isolation and characterisation of genomic and cDNA clones of human erythropoietin. Nature 1985;313:806–810.
14 Johnson CS, Cook CA, Furmanski P: In vivo suppression of erythropoiesis by tumor necrosis factor-alpha: Reversal with exogenous erythropoietin. Exp Hematol 1990; 18:109–113.
15 Koury ST, Bondurant MC, Koury MJ: Localization of cells containing erythropoietin messenger RNA in the kidneys on anemic mice using in situ hybridization. Blood 1987;70(suppl 1):176A.
16 Law ML, Cai GY, Lin FK, Wei A, Huang JH, Hartz H, Morse CH, Lin C, Jones C, Kao FT: Chromosomal assignment of the human erythropoietin gene and its DNA polymorphism. Proc Natl Acad Sci USA 1986;83:6920–6924.
17 Levine RL, Englard A, McKinley GF, Lubin W, Abels RI: The efficacy and lack of toxicity of escalating doses of recombinant erythropoietin in anemic AIDS patients on Zidovudine. Blood 1989:74(suppl 1):15A.

18 Lim VS, DeGowin RL, Zavala D, Kirchner PT, Abels R, Perry P, Fangman J: Recombinant human erythropoietin treatment in pre-dialysis patients. Ann Intern Med 1989;110:108–114.
19 Lin FH, Suggs S, Lin CH, Browne JK, Smalling R, Egrie JC, Chen KK, Fox GM, Martin F, Stabinsky Z, Badrawi SM, Lai PH, Goldwasser E: Cloning and expression of the human erythropoietin gene. Proc Natl Acad Sci USA 1985;82:7580–7584.
20 Ludwig H, Fritz E, Kotzmann H, Höcker P, Gisslinger H, Barnas U: Erythropoietin treatment for chronic anemia of malignancy. Blood 1989;74(suppl 1):16A.
21 Means RT, Olsen NJ, Krantz SB, Graber SE, Dessypris EN, Stone WJ, O'Neil V, Pincus TP: Treatment of the anemia of rheumatoid arthritis with recombinant human erythropoietin: Clinical and in vitro studies. Arthritis Rheum 1989;32:638–642.
22 Miller C, Jones R, Piantadosi S, Abeloff M, Spivak J: Decreased erythropoietin response associated with the anemia of malignancy. Proc ASCO 1989;8:709A.
23 Oster W, Herrmann F, Cicco A, Gamm H, Zeile G, Brune T, Lindemann A, Schulz G, Mertelsmann R: Erythropoietin prevents chemotherapy-induced anemia: Case report. Blut 1990;60:88–92.
24 Oster W, Herrmann F, Gamm H, Zeile G, Lindemann A, Müller H-G, Brune T, Kraemer H-P, Mertelsmann R: Erythropoietin for the treatment of anemia of malignancy associated with neoplastic bone marrow infiltration. J Clin Oncol 1990;8:956–962.
25 Oster W, Herrmann F, Lindemann A, Mertelsmann R: Experimental and clinical evaluation of erythropoietin; in Habenicht A (ed): Growth Factors, Differentiation Factors, and Cytokines. Stuttgart, Springer, 1990, pp 232–242.
26 Sawyer ST, Hosoi T, Krantz SB: Structure of the erythropoietin receptor. Blood 1988;72:133A.
27 Schuster SJ, Wilson J, Erslev AJ, Caro J: Physiologic regulation and tissue localization of renal erythropoietin mRNA. Blood 1986;68(suppl 1):179A.
28 Tischler HJ, Leblanc S, Brune H, Welte K, Poliwoda H, Link H: Recombinant human erythropoietin in treatment of pure red cell anemia. Blut 1989;59:340A.
29 Vellenga E, De Wolf JTM, Halie MR: Recombinant erythropoietin failed to correct anemia in Fanconi syndrome. Leukemia 1989;5:858.
30 Winearls CG, Oliver DO, Pippard MJ, Reid C, Downing MR, Cotes PM: Effect of human erythropoietin derived from recombinant DNA on the anemia of patients maintained by chronic hemodialysis. Lancet 1986;ii:1175–1178.
31 Zanjani ED, Ascensao JL, McGlave PB, Banisadre M, Ash RC: Studies on the liver to kidney switch of erythropoietin production. J Clin Invest 1981;67:1183–1188.

Wolfgang Oster, MD, Behringwerke AG, PO Box 11 40,
Bunsenstrasse 1, D-3550 Marburg/Lahn (FRG)

Improved Function of B Lymphocytes in Dialysis Patients Treated by Recombinant Human Erythropoietin

Leszek Paczek, Roland M. Schaefer, August Heidland[1]

Department of Internal Medicine, Division of Nephrology,
University of Würzburg, FRG

Studies addressing the influence of recombinant human erythropoietin (rhEPO) on the immune system are up to now limited and inconsistent. Pfäffl et al. [1] reported on a decrement of the suppressor cell subpopulation, an increased helper/suppressor ratio, and a higher score in delayed cutaneous hypersensitivity testing in dialysis patients treated with rhEPO. On the other hand, the blastogenic response to phytohemagglutinin (PHA), concanavalin A (Con A), and in mixed lymphocyte cultures (MLC) was subnormal in uremic patients before and after correction of their anemia [2, 3]. Regarding humoral immunity, the absolute number of circulating B cells was reduced during the initial period of rhEPO therapy, with a return to baseline levels after 18 weeks of treatment [1].

The present investigation on immunoglobulin production by cultured lymphocytes from uremic patients receiving rhEPO was performed to gain further understanding of the interaction between rhEPO and immunocompetent cells.

Methods

Patients and Subjects

Group I consisted of 15 long-term hemodialysis patients (8 males, 7 females) with a mean age of 56 ± 13 years, being dialyzed for 40 months (range 9–66). These patients had received an average dose of 80 U/kg rhEPO/week for 15 ± 5 months. Under this regimen, hemoglobin levels had increased from 7.1 ± 0.8 to 9.3 ± 1.1 g/100 ml. A subgroup of 5 patients was examined before and after 8 and 16 weeks of therapy with rhEPO.

---

[1] We would like to thank M. Schott for his skilful technical assistance.

Group II consisted of 14 hemodialysis patients (7 males, 7 females) with a mean age of 59 ± 12 years, being dialyzed for 41 months (range 12–72). These patients did not receive rhEPO, but had hemoglobin values (9.0 ± 1.7 g/100 ml) in the range of group I receiving rhEPO.

Group III was a group of healthy controls consisting of 8 males and 7 females with a mean age of 55 ± 5 years. They had normal renal function with a serum creatinine level of 1.1 ± 0.3 mg/100 ml and their hemoglobin was 14.1 ± 0.8 g/100 ml.

Immunoglobulin Production by PBMC

Peripheral venous blood was taken and anticoagulated with 20 U/ml of preservative-free heparin (Sigma, Munich, FRG). The blood was diluted twice with Hanks' balanced salt solution and peripheral blood mononuclear cells (PBMC) were separated on a Histopaque solution gradient (Sigma). Cells were washed 3 times and resuspended in culture medium in a concentration of 2×10^6/ml. The medium contained RPMI 1640, L-glutamine (4 mM), Hepes (25 mM), penicillin (100 U/ml), streptomycin (100 µg/ml), nystatin (250 U/ml) and was supplemented with 10% heat-inactivated fetal calf serum (Gibco, Grand Island, N.Y., USA). For stimulation of PBMC, pokeweed mitogen (PWM; Gibco) in a final dilution of 1:200 or *Staphylococcus aureus* Cowan strain I (SAC; Calbiochem, Frankfurt, FRG) in a dilution of 1:4,000 was used. Cells were maintained in a final volume of 500 µl in polystyrole tubes (Falcon, Heidelberg, FRG) at 37 °C in 5% CO_2 in a humidified incubator. After 7 days of culture, supernatants were harvested and stored at –80 °C until analyzed.

The concentrations of IgA, IgG, and IgM were measured by enzyme-linked immunosorbent assays (ELISA) according to the method of Lems-Van Kan et al. [4]. The assays were performed in flat-bottomed microplates (Linbro; McLean, Va., USA). Goat antibodies against human IgA, IgG, and IgM were purchased from Zymed, San Francisco, Calif., USA, and o-phenylenediamine dihydrochloride (OPD), the substrate for horseradish peroxidase, from Sigma. The optical density of each well was estimated by an ELISA reader obtained from Behring, Marburg, FRG.

Statistical Analysis

All results were expressed as means ± SEM. Statistical evaluation was performed by the Student's t test for paired or unpaired data as appropriate.

Results

Immunoglobulin Production by Uremic and Healthy PBMC

In comparison to healthy controls, basal formation of both IgA and IgG were decreased in PBMC from uremic patients. IgM production was also lower in dialysis patients, but this difference was not significant. PWM, a T cell-dependent stimulus, caused a much higher response in B lymphocytes in terms of IgA and IgG secretion from healthy volunteers as compared with uremic patients. The same held true for stimulation of B cells by SAC, which represents a T cell-independent stimulant. Again PBMC from healthy controls produced more immunoglobulins than PBMC derived from uremics. In

Table 1. Immunoglobulin production by PBMC from uremics and healthy controls (ng/ml in supernatants)

		Controls (Hct 43%) n = 14	Uremics (Hct 21%) n = 14	Uremics (Hct 30%) n = 13
IgA	Basal	589±78[a]	241±44	280±69
	PWM	3,017±433[a]	1,035±330	1,506±328
	SAC	1,317±272[a]	375±80	371±62
IgG	Basal	674±94[a]	392±51	308±73
	PWM	3,596±419[a]	1,764±304	1,910±452
	SAC	3,063±411[a]	957±361	1,329±433
IgM	Basal	64±34	47±6	35±7
	PWM	1,354±141	939±182	1,156±382
	SAC	810±186	783±317	889±181

Data are given as means ± SEM.
[a] $p < 0.01$ for controls versus uremics.

terms of B cell function, both basal and stimulated, there were no differences between severe or moderately anemic hemodialysis patients (table 1).

Immunoglobulin Production by PBMC from Uremics Receiving rhEPO

Immunoglobulin production by PBMC from 8 dialysis patients was followed before and after 8 weeks of rhEPO. The hematocrit had risen from 21 ± 1 to 33 ± 1%. With stimulation of PBMC by PWM, IgG secretion rose significantly during rhEPO administration by 67%, while IgA formation increased by 21% ($p < 0.05$). The production of IgM by PBMC from uremics was not influenced by rhEPO treatment (fig. 1). Measurements of immunoglobulin production in 5 patients after 16 weeks of rhEPO treatment revealed a decline of IgA and IgG formation towards baseline levels.

In vitro Effect of rhEPO on Healthy PBMC

To test if this rise in immunoglobulin production was a specific effect of rhEPO on B cells, PBMC from healthy donors (n = 14) were incubated with the hormone (2 U/ml) for 7 days. Basal antibody formation was not changed by the presence of rhEPO in the culture medium. As can be seen in figure 2, SAC-stimulated immunoglobulin production was generally higher in cultures containing rhEPO. Formation of IgA increased by 15% ($p < 0.05$), IgM by 30% (NS), and IgG by 20% ($p < 0.05$).

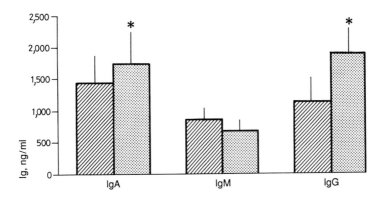

Fig. 1. Effect of rhEPO (8 weeks) on PWM-stimulated antibody formation. Immunoglobulins (Ig) were determined in supernatants from PBMC cultures (7 days) obtained from 5 hemodialysis patients and are given in ng/ml. Data are expressed as means ± SEM. *p < 0.05 for values before (▨) versus during (▩) rhEPO.

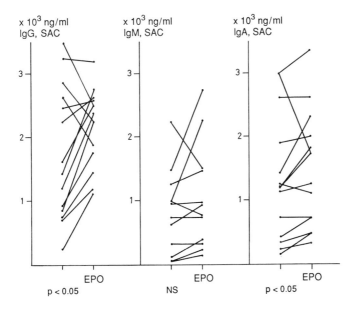

Fig. 2. In vitro effect of rhEPO on SAC-stimulated PBMC from healthy subjects (n = 14). Data are expressed as the individual increments of immunoglobulin production in cultures supplemented with rhEPO (2 U/ml) as compared to those without the hormone. p < 0.05 for cultures with versus without rhEPO.

Discussion

The present study demonstrates clearly that immunoglobulin production by B lymphocytes (basal and stimulated) is reduced in the state of end-stage renal disease. It was shown that this defect in B cell function was not dependent on the degree of anemia, since basal and stimulated antibody formation by lymphocytes was comparable in uremics with severe or a more moderate type of anemia (table 1).

Eight weeks after the onset of the antianemic therapy with rhEPO, the stimulated production of IgA and IgG by B cells was increased without reaching the range of healthy subjects (fig. 1). After 16 weeks of treatment, however, the stimulatory effect of rhEPO on antibody production had abated.

To test whether enhanced antibody secretion was due to a direct interaction of rhEPO with PBMC, the hormone was added (2 U/ml) to the culture medium. The formation of antibodies was unchanged in PBMC under basal conditions, but was enhanced in SAC-stimulated cells. This observation strongly suggests that rhEPO exerts a direct effect on cells involved in antibody production (B cells) or regulation of these processes (T helper cells and monocytes).

It could be argued that the employed dose of rhEPO was rather above the physiological range (0–25 mU/ml). However, at least shortly after the intravenous administration of rhEPO such high levels can be encountered in the serum of patients [5]. What is more, Rich et al. [6] were able to demonstrate that EPO is not only produced by the kidney or liver, but also by macrophages. Based upon their observations, these authors postulate a pivotal role for the macrophage in the regulation of erythropoiesis in the bone marrow, a hypothesis which would imply rather high local concentrations of EPO in the bone marrow, which would be similar to the situation in our in vitro experiments.

In terms of direct effects of rhEPO on lymphocytes and/or monocytes a second question arises, namely whether there are specific receptors for EPO on these cells. This particular problem was addressed by Krantz et al. [7] who followed the binding of labeled rhEPO. According to these authors, there was no specific binding to liver, kidney, brain or lung, nor to B and T lymphocytes and monocytes. Koury et al. [8], however, described binding of labeled EPO by the placenta of pregnant mice.

In summary, substitution therapy with rhEPO in long-term hemodialysis patients resulted in improved antibody formation by B lymphocytes. It seems that this was not due to correction of anemia, but was rather a specific effect of the hormone. This stimulatory effect, however, did not prevail and after 16 weeks of rhEPO treatment, the antibody production by B cells had returned to what was observed prior to the antianemic therapy. The clinical implications therefrom are twofold: The immunological response to a grafted kidney could be more pronounced in a recipient being in the early phase of rhEPO therapy. In fact, there is one report in the literature [9] that graft function during the first 3 weeks after transplantation was worse in patients who were on rhEPO as compared with untreated patients. On the other hand, a state of improved immunological competence might be of benefit in cases of acute or chronic infection.

References

1 Pfäffl W, Gross HJ, Neumeier D, Nattermann U, Samtleben W, Gurland HJ: Lymphocyte subsets and delayed cutaneous hypersensitivity in hemodialysis patients receiving recombinant human erythropoietin. Contrib Nephrol. Basel, Karger, 1988, vol 66, pp 195–204.
2 Fehrmann I, Barany P, Bergström J: Immunological studies in patients treated with human erythropoietin. Abstr EDTA, Gothenburg, 1989, p 199.
3 Modai D, Berman S, Averbukh Z, Cohn M, Weissgarten J, Golik A, Cohen N: Recombinant human erythropoietin does not alter various immunological parameters in uremic patients. Abstr EDTA, Gothenburg, 1989, p 208.
4 Lems-Van Kan P, Verspaget HW, Pena AS: ELISA assay for quantitative measurement of human immunoglobulins IgA, IgG, and IgM in nanograms. J Immunol Methods 1983;57:51–57.
5 Bolaert J, Schurgers M, Matthys E, Daneels R, DeCre M, Bogaert M: Recombinant human erythropoietin pharmacokinetics in CAPD patients: comparison of the intravenous, subcutaneous, and intraperitoneal routes. Nephrol Dial Transplant 1989;3:493.
6 Rich IN, Heit W, Kubanek B: Extrarenal erythropoietin production by macrophages. Blood 1982;60:1007–1018.
7 Krantz SB, Saweyer ST, Sawada KI: The role of erythropoietin in erythroid cell differentiation. Contrib Nephrol. Basel, Karger, 1988, vol 66, pp 25–37.
8 Koury MJ, Bondurant MC, Graber SE: Erythropoietin messenger RNA levels in developing mice and transfer of ^{125}I-erythropoietin by the placenta. J Clin Invest 1988;82:154–159.
9 Wahlberg J, Jacobson J, Odlind B, Tufveson G, Wikström B: Haemodilution in renal transplantation in patients on erythropoietin. Lancet 1988;ii:1418.

Roland M. Schaefer, MD, Medizinische Universitätsklinik,
Josef-Schneider-Strasse 2, D–8700 Würzburg (FRG)

Influence of Erythropoietin Treatment on Glucose Tolerance, Insulin, Glucagon, Gastrin and Pancreatic Polypeptide Secretion in Haemodialyzed Patients with End-Stage Renal Failure[1]

Franciszek Kokot, Andrzej Więcek, Władysław Grzeszczak, Mariusz Klin, Ewa Żukowska-Szczechowska

Department of Nephrology, Silesian School of Medicine, Katowice, Poland

Erythropoietin (EPO) deficiency is an undoubted factor participating in the pathogenesis of anaemia in patients with chronic uraemia. On the other side it is well known that other hormones (thyroid hormones, gonadal hormones, cortisol, etc.) are modulators of erythropoiesis even under physiological conditions. From data obtained during the last 3 years [1–11] it seems that some clinical signs and symptoms such as increased well-being, libido and potency, increased appetite, exacerbation of pre-existing elevated blood pressure and improvement of physical and mental activity in EPO-treated patients are not only related to amelioration of anaemia.

In the present study we aimed to assess the influence of short-term treatment with EPO on secretion of selected gastrointestinal and pancreatic hormones, which are involved in several metabolic pathways and regulation of the digestive function of the gastrointestinal tract.

Material and Methods

Five haemodialyzed patients (aged 39–57 years, mean duration of haemodialysis treatment 48.2 ± 23.0 months, mean predialysis creatinine level 1,208 ± 46.5 μmol) were examined before and after 3 months of EPO treatment (EPO group). Results

---

[1] Supported in part by the Polish Ministry of Health and Welfare, MZ XIII.

obtained in this group under basic conditions were compared with those of 6 haemodialyzed patients (designated as the non-EPO group, aged 30–50 years, mean duration of dialysis treatment 73.0 ± 21.3 months, mean predialysis plasma creatinine level 929 ± 46.0 µmol/l) in whom a similar blood haemoglobin level (6.7 ± 0.3 mmol/l) and Hct value (34.4 ± 1.4%) were found as in EPO-treated patients after completed therapy. Patients of the EPO group were treated with EPO (Cilag AG, Schaffhausen, Switzerland) for 3 months. The average dose of EPO was 75 U/kg b.w. 3 times/week as a bolus injection given after completion of a dialysis session. The control group consisted of 10 healthy subjects (aged 23–56 years, mean creatinine level 93.2 ± 2.4 µmol/l).

None of the examined subjects were overweight or suffering from diabetes. The subjects were given a normalized diet containing 60 g protein, 70 g fat and 325 g carbohydrate for at least 7 days before a test meal was performed. In brief, the test meal was done according to the following protocol: In the early morning after withdrawing a blood sample (0-min sample) the examined subjects ingested a test meal which contained 37.0 g protein, 69.0 g fat and 108.0 g carbohydrate. Further blood samples were withdrawn 15, 30, 45, 60, 90, 120, 150, 180 and 240 min after ingestion of the test meal. In patients on EPO treatment a test meal was performed twice, i.e. before and after 3 months of EPO theray. In these patients blood samples were also withdrawn 3 months after discontinued EPO therapy but only under basic conditions.

In all blood samples blood glucose levels and the following hormonal parameters were assessed: immunoreactive insulin (IRI), glucagon (Glu), gastrin (Ga) and pancreatic polypeptide (PP) using radioimmunological methods as specified previously [12, 13]. All the other parameters (Hct, Hb, creatinine) were estimated by routine methods [14].

Statistical evaluation of results was done using the Students t test for paired or unpaired variables, comparing each time only two of the examined groups respectively [15].

Results

Haematological Indices

In patients on EPO therapy the pre- and posttreatment haemoglobin levels were 4.47 ± 0.2 and 6.67 ± 0.15 mmol/l, respectively, while the haematocrit (Hct) value rose from 23.0 ± 0.9 to 34.6 ± 0.7%. Three months after discontinuation of EPO therapy, the respective Hb and Hct values were 5.3 ± 0.2 mmol/l and 27.6 ± 1.0%. In healthy subjects the Hb concentration was 8.4 ± 0.1 mmol/l and the Hct 43.3 ± 1.1%, respectively.

Glucose

As can be seen in figure 1, pretreatment blood glucose curve was markedly higher in patients on EPO than in healthy subjects. After 3 months of EPO treatment the glycaemic curve after ingestion of a test meal was superposable with that in normals. The pre- and posttreatment area under the curve (AUC) was 27.0 ± 2.54 and 19.0 ± 1.06 mmol·l^{-1}·h respectively

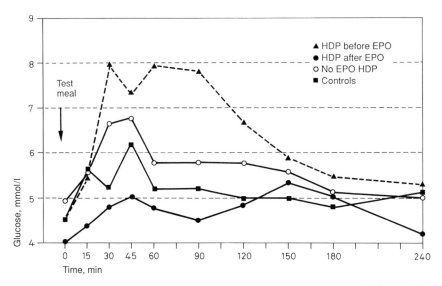

Fig. 1. Baseline plasma glucose levels and the response to a test meal in haemodialyzed patients (HDP) before and after 3 months of EPO treatment, in haemodialyzed patients not treated by EPO (No EPO HDP) and in healthy subjects.

($p < 0.02$) while in healthy subjects 20.51 ± 1.31 mmol·l^{-1}·h ($p < 0.005$ vs. pretreatment values in EPO patients). After 3 months of discontinued EPO therapy the blood glucose level was 4.44 ± 0.1 mmol/l. In uraemic patients not treated with EPO the fasting blood glucose level was 4.92 ± 0.09 mmol/l. As shown in figure 1, in patients of the non-EPO group the glucose curve after ingestion of a test meal was nearly superposable with that in healthy subjects, and located between the respective pre- and posttreatment curves of patients treated with EPO. The AUC of this group (22.43 ± 0.42 mmol·l^{-1}·h) was of the same magnitude as in normals but significantly ($p < 0.05$) lower than the pretreatment one and significantly higher than the posttreatment one of the EPO group.

Insulin

Basal plasma levels of IRI before EPO treatment (20.0 ± 2.2 μU/ml) were significantly ($p < 0.01$) higher than in healthy subjects (13.0 ± 1.0 μU/ml). After 3 months on EPO and 3 months after discontinued EPO administration, IRI levels were significantly ($p < 0.01$) higher (37.0 ± 6.4 and 40.8 ± 7.3 μU/ml, respectively) than pretreatment levels and not signifi-

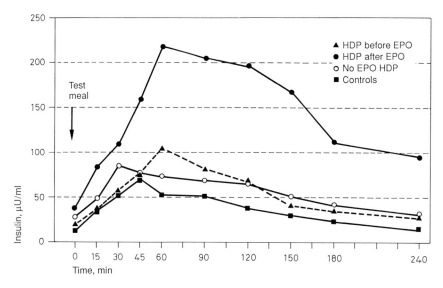

Fig. 2. Baseline plasma insulin levels and the response to a test meal in haemodialyzed patients (HDP) before and after 3 months of EPO treatment, in haemodialyzed patients not treated by EPO (No EPO HDP) and in healthy subjects.

cantly higher than in the non-EPO group (29.17 ± 1.19 µU/ml). As can be seen in figure 2, the pretreatment IRI response to a test meal was moderately ($p < 0.05$) greater (AUC 234.4 ± 71 µU·ml^{-1}·h) than in normals (AUC 144.1 ± 24 µU·ml^{-1}·h), but significantly ($p < 0.005$) smaller than after EPO treatment (AUC 589.5 ± 17.1 µU·ml^{-1}·h). In patients of the non-EPO group the AUC for IRI was of the same magnitude (223 ± 13 µU·ml^{-1}·h) as the pretreatment one of the EPO group, but significantly ($p < 0.02$) lower than the respective posttreatment value.

Glucagon

Pretreatment basal plasma Glu levels (258 ± 71 pg/ml) were significantly ($p < 0.05$) higher than those after EPO treatment (101 ± 19 pg/ml) and significantly ($p < 0.05$) higher than in healthy subjects (120 ± 26 pg/ml). After 3 months of discontinued EPO therapy, basal plasma levels of Glu again increased to pretreatment levels (275 ± 28 pg/ml). Basal plasma levels of Glu in uraemic patients of the non-EPO group were lower (177 ± 14 pg/ml) than pretreatment levels but moderately higher than posttreatment levels of the EPO group. As can be seen in figure 3, the Glu response to a test meal

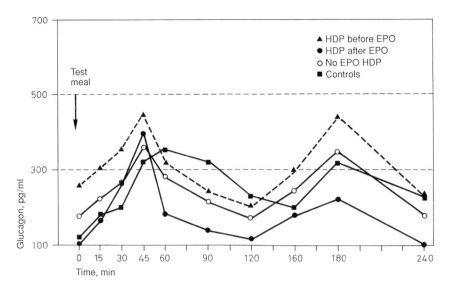

Fig. 3. Baseline plasma glucagon levels and the response to a test meal in haemodialyzed patients (HDP) before and after 3 months of EPO treatment, in haemodialyzed patients not treated by EPO (No EPO HDP) and in healthy subjects.

before EPO treatment was significantly ($p < 0.02$) higher (AUC $1{,}251 \pm 171$ pg·ml^{-1}·h) than after EPO therapy (AUC 671 ± 88 pg·ml^{-1}·h) but of the same magnitude as in healthy subjects (AUC $1{,}133 \pm 127$ pg·ml^{-1}·h). In patients of the non-EPO group the AUC for glucagon ($1{,}010 \pm 70$ pg·ml^{-1}·h) was of the same magnitude as in healthy subjects and moderately smaller than the pretreatment one in EPO-treated patients.

Gastrin

Pretreatment basal plasma Ga levels in patients of the EPO group (125 ± 19 pg/ml) were significantly higher ($p < 0.001$) than in normals (38 ± 6 pg/ml) but significantly ($p < 0.05$) lower than in uraemic patients of the non-EPO group (265 ± 16 pg/ml). After 3 months of EPO therapy, plasma Ga levels declined significantly ($p < 0.001$) and were in the normal range (48 ± 16 pg/ml). Three months after discontinued EPO therapy, plasma Ga levels again raised to pretreatment ones (120 ± 9 pg/ml). As shown in figure 4, the pretreatment response of plasma Ga to a test meal in patients of the EPO group was significantly ($p < 0.005$) higher (AUC 963 ± 179 pg·ml^{-1}·h) than in normals (AUC 226 ± 48 pg·ml^{-1}·h) and was

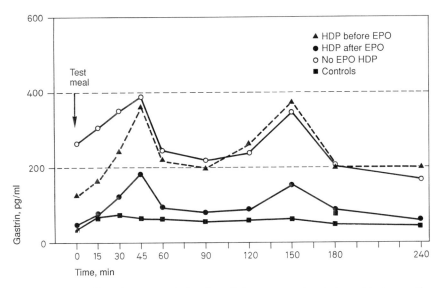

Fig. 4. Baseline plasma gastrin levels and the response to a test meal in haemodialyzed patients (HDP) before and after 3 months of EPO treatment, in haemodialyzed patients not treated by EPO (No EPO HDP) and in healthy subjects.

significantly ($p < 0.02$) suppressed after EPO therapy (AUC 392 ± 93 pg·ml^{-1}·h). In patients of the non-EPO group the AUC for Ga (1,031 ± 62 pg·ml^{-1}·h) was of similar magnitude as the respective pretreatment value of the EPO group but significantly ($p < 0.001$) higher than the posttreatment value of EPO-treated patients and the value found in healthy subjects.

Pancreatic Polypeptide

Pretreatment basal plasma pancreatic polypeptid (PP) levels in patients of the EPO group were significantly ($p < 0.001$) higher (734 ± 92 pg/ml) than in normals (80 ± 12 pg/ml) and declined moderately after EPO therapy (526 ± 93 pg/ml; $p < 0.02$). Three months after discontinued EPO therapy there was a further decline of basal plasma PP levels to 250 ± 34 pg/ml. In uraemic patients of the non-EPO group, basal plasma PP levels were of the same magnitude (682 ± 32 pg/ml) as pretreatment values in the EPO group.

As can be seen in figure 5, the pretreatment response of plasma PP to a test meal was significantly ($p < 0.001$) more marked (AUC 3,658 ± 577 pg·ml^{-1}·h) than in healthy subjects (AUC 1,663 ± 164 pg·ml^{-1}·h) and was moderately suppressed ($p < 0.05$) after EPO treatment (AUC 2,989 ± 473

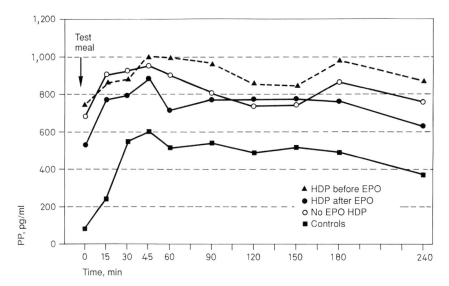

Fig. 5. Baseline plasma pancreatic polypeptide levels and the response to a test meal in haemodialyzed patients (HDP) before and after 3 months of EPO treatment, in haemodialyzed patients not treated by EPO (No EPO HDP) and in healthy subjects.

$pg \cdot ml^{-1} \cdot h$). In patients of the non-EPO group the response of plasma PP to a test meal was similar as in EPO-treated patients but significantly ($p < 0.001$) more marked than in healthy subjects.

Discussion

Hyperinsulinism and elevated plasma levels of glucagon, gastrin and pancreatic polypeptide are well known endocrine abnormalities in haemodialyzed and nonhaemodialyzed uraemic patients [for references, see 16]. These hormonal abnormalities are regarded as a consequence of decreased capability of the residual renal parenchyma to eliminate or degrade the above mentioned polypeptide hormones and/or of their enhanced secretion by the respective endocrine organ. Increased secretion of the hormones in uraemic patients may be due to decreased binding of hormones at the receptor site (caused by 'uraemic toxins'), suppressed intracellular amplification of the hormonal signal after internalization or presence of factors antagonizing biological effects of the respective hormone. As is well known, kidneys are an important organ where all the above-mentioned hormones are eliminated or degraded. On the other side it was shown that EPO treatment does not influence the excretory function of residual renal parenchyma in uraemic patients [17]. Thus it seems highly unlikely that altered biodegradation or elimination of hormones by residual kidneys is the only or main cause of hormonal alterations found in EPO-treated patients. This statement seems to be supported by the fact that after EPO therapy both basal plasma levels of insulin and the response of plasma insulin to a test meal were increased while those of glucagon, gastrin and pancreatic significantly or moderately suppressed. From these results it follows that EPO-induced hormonal alterations are rather caused by altered secretion than by altered degradation of these hormones. As both basal- and test-meal-stimulated plasma levels of insulin, glucagon, gastrin and pancreatic polypeptide in patients after EPO treatment differed from those found in uraemic patients with similar values of Hct and Hb as patients after EPO therapy, it seems unlikely that improvement of anaemia per se was the only cause of the observed hormonal alterations.

Results obtained in this paper prove improvement of carbohydrate tolerance in EPO-treated haemodialyzed uraemic patients. This improvement could be due to enhanced insulin release as well as to suppression of glucagon secretion. As insulin secretion is markedly dependent upon incretin factors released from the gastrointestinal tract, the contribution of altered function of the enteroinsular axis in the pathogenesis of improved carbohydrate tolerance in EPO-treated patients remains to be elucidated.

Calculation of the so-called insulin:glucose index (by dividing the AUC for insulin by the AUC for glucose) revealed significant increase of this index from 8.7 to 31 in patients after treatment with EPO. In patients of the non-EPO group and in healthy subjects the respective values were 9.95 and 6.85. These data prove the presence of increased insulin resistance in EPO-treated patients and suggest that improvement of carbohydrate tolerance results mainly at the expense of increased insulin secretion.

Results presented in this paper suggest that EPO treatment shows a profound effect on function of endocrine organs which in turn may be involved in the pathogenesis of improved well-being and physical and mental activity in EPO-treated patients. Some EPO-induced endocrine alterations seem to be confined only to the period of EPO administration (gastrin and glucagon secretion) while others are still present 3 months after EPO withdrawal (insulin and PP secretion). It cannot be excluded that long-term EPO therapy may be followed by adaptive and compensatory mechanisms, which in turn may induce endocrine alterations different from those reported in this paper. Thus, further studies on function of endocrine organs in long-term EPO-treated patients are mandatory.

References

1. Winearls CG, Oliver DO, Pippard MJ, Reid C, Downing MR, Cotes PM: Effect of human erythropoietin derived from recombinant DNA on the anaemia of patients maintained by chronic haemodialysis. Lancet 1986;ii:1175–1178.
2. Eschbach JW, Egrie JC, Downing MR, Browne JK, Adamson JW: Correction of the anaemia of end-stage renal disease with recombinant human erythropoietin. New Engl J Med 1987;316:73–78.
3. Casati S, Passerini P, Campise MR, Graziani G, Cesana B, Perisie M, Ponticelli C: Benefits and risks of protracted treatment with human recombinant erythropoietin in patients having haemodialysis. Br Med J 1987;295:1017–1020.
4. Bommer J, Alexiou C, Müller-Bühl E, Eifert J, Ritz E: Recombinant human erythropoietin therapy in haemodialysis patients – dose determination and clinical experience. Nephrol Dial Transplant 1987;2:238–242.
5. Mayer G, Thum J, Cade EM, Stumvoll HK, Graf H: Working capacity is increased following recombinant human erythropoietin treatment. Kidney Int 1988;34:525–528.
6. Schaefer RM, Leschke M, Strauer BE, Heidland A: Blood rheology and hypertension in hemodialysis patients treated with erythropoietin. Am J Nephrol 1988;8:449–453.
7. Grützmacher P, Bergmann M, Weinreich T, Nattermann U, Reimers E, Pollok M: Beneficial and adverse effects of correction of anaemia by recombinant human erythropoietin in patients on maintenance haemodialysis. Contrib Nephrol. Basel, Karger, 1988, vol 66, pp 104–113.
8. Böcker A, Reimers E, Nonnast-Daniel B, Kühn K, Koch KM, Scigalla P, Braumann KM, Brunkhorst R, Böning D: Effect of erythropoietin treatment on O_2 affinity and performance in patients with renal anemia. Contrib Nephrol. Basel, Karger, 1988, vol 66, pp 165–175.
9. Schaefer RM, Kokot F, Wernze H, Geiger H, Heidland A: Improved sexual function in hemodialysis patients on recombinant erythropoietin: A possible role of prolactin. Clin Nephrol 1989;31:1–5.
10. Schaefer RM, Hörl WH, Massry SG. Treatment of renal anemia with recombinant human erythropoietin. Am J Nephrol 1989;9:353–362.
11. Sundal E, Kaeser U: Correction of anaemia of chronic renal failure with recombinant human erythropoietin: Safety and efficacy of one year's treatment in a European Multicenter Study of 150 haemodialysis-dependent patients. Nephrol Dial Transplant 1989;4:979–987.
12. Grzeszczak W, Kokot F, Duława J: Effect of naloxone administration on endocrine abnormalities in chronic renal failure. Am J Nephrol 1987;7:93–100.
13. Kokot F, Stupnicki R: Radioimmunological and Radiocompetitive Methods Used in Clinical Medicine. Warsaw, National Medical Publisher, 1985.
14. Kokot F: Laboratory Methods Used in Clinical Practice. Warsaw, National Medical Publisher, 1969.

15 Armitage P: Statistical Methods in Medical Research. Warsaw, National Medical Publisher, 1978.
16 Kokot F, Więcek A: Endocrine changes in chronic dialysis patients; in Maher JF (ed): Replacement of Renal Function by Dialysis. Dordrecht, Kluwer Academic Publishers, 1989, pp 953–971.
17 Onoyama K, Kumagai H, Takeda K, Shimamatsu K, Fujishima M: Effects of human recombinant erythropoietin on anemia, blood pressure and renal function in predialysis renal failure patients. Nephrol Dial Transplant 1989;4:966–970.

Prof. Dr. hab. med. Franciszek Kokot, Department of Nephrology,
Silesian School of Medicine, Francuska 20/24 Str. PL–40-027 Katowice (Poland)

Changes in Red Blood Cell Volume under Recombinant Human Erythropoietin Therapy

Reinhard Schmidt[a], Dietmar Lerche[b], Erhard Dörp[a], Roland Winkler[a], Horst Klinkmann[a]

[a] Department of Internal Medicine, University of Rostock, and
[b] Institute of Medical Physics and Biophysics of the Charité, Humboldt University, Berlin, GDR

Various authors have reported about changes in blood rheology under recombinant human erythropoietin (rhEPO) therapy [5, 9, 11, 12]. The increase of apparent whole blood viscosity has been held responsible for deteriorating preexisting hypertension [1, 11]. In a former study [6] it was shown that membrane mechanics and deformability of the single red blood cell (RBC) could be improved remarkably under rhEPO therapy. From these results the question arose whether the improved rheological RBC function is associated with a prolonged life span and other changes of the RBC. Continuing the studies on the function of RBC under rhEPO treatment, investigation about the life span and the cell volume of RBC have been undertaken.

Material and Methods

Twenty-four patients (mean age 43.1 ± 12.6 years) with end-stage renal failure, who were on regular dialysis 3 times weekly, were investigated. The frequency of blood transfusion was averaged by 9.1 transfusion units per 6 months. Initial hematocrit (Hct) was 21.3 ± 3.5%. According to the study protocol, a rhEPO dosage of 80 IU/kg body weight was administered until the target Hct of 30–35% was reached. After reaching the target Hct an individual maintenance dosage was given. Hematological, biochemical and clinical data were recorded following the given protocol. Investigations regarding RBC life span and RBC volume were performed before the study and 6 months after starting rhEPO therapy in the same individuals.

RBC life span: 9 cm^3 blood drawn from the patient was mixed with 1 cm^3 natrium citricum, labeled with ^{51}Cr and retransfused. Activity of the blood samples taken every other day was measured and RBC life span was calculated.

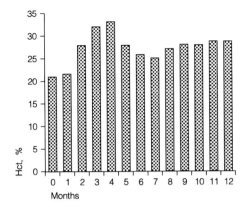

Fig. 1. Changes of Hct under rhEPO therapy.

RBC filterability: The whole cell deformation of RBC was measured by a filtrometer working by the gravity principle. Patients' RBC were washed and resuspended to a Hct of 60%. The pores of the filters used were 20–40 µm in size. Initial maximum pressure of 300 Pa was applied. The time necessary to pass 0.5 ml of the RBC suspension through the filter divided by the time necessary to permeate for the suspension medium alone gives the filterability index.

RBC volume measurement by micropipette aspiration: Under microscopical observation a portion of the RBC was sucked under pressure control into the cylindrical part of a glass microcapillary with an interior diameter of 1.1–1.6 µm. The part of the RBC outside the capillary is ball-shaped and the part inside the capillary is a cylinder. The volume of these geometrical figures can be calculated. All lengths, e.g. capillary radius and length of RBC tongue inside the capillary, were measured digitally with an electronic TV line analyzer. From each sample at least 250 cells were investigated [7, 8].

For statistical analysis, mean values and standard deviation were calculated and significance was tested by the paired Student's t test. In cases of small groups, statistical analysis has not been undertaken.

Results

The target Hct in the 24 patients evaluated was reached after 10 weeks (Hct 31.6 ± 2.9%). In order to maintain the Hct in this range, an average rhEPO dose of 41.4 ± 19.8 IU/kg body weight was necessary. Figures 1–3 show the changes of Hct, RBC count and absolute numbers of reticulocytes over an observation period of 12 months.

Since the investigations of RBC life span, RBC filterability and RBC volume were done before and 6 months after rhEPO therapy, table 1 shows

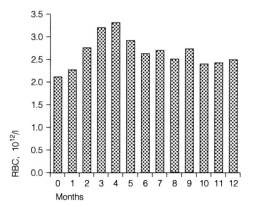

Fig. 2. Changes of RBC count under rhEPO therapy.

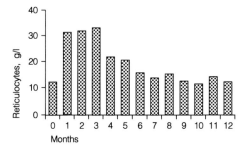

Fig. 3. Changes of the absolute number of reticulocytes under rhEPO therapy.

Table 1. Hematological data before and 6 months after rhEPO therapy (n = 24)

	Before rhEPO therapy		After 6 months of rhEPO therapy		p
	mean ± SD	%	mean ± SD	%	
Hematocrit, %	21.3 ± 3.5	100	26.4 ± 4.1	123.9	< 0.05
Erythrocytes, 10^{12}/l	2.31 ± 0.32	100	2.72 ± 0.46	117.7	< 0.05
Mean cell corpuscular volume	92.2	100	97.1	105.3	
Reticulocytes, g/l	12.4 ± 2.4	100	16.6 ± 3.7	133.9	< 0.05

Table 2. RBC life span before and 6 months after rhEPO therapy (n = 6; mean ± SD)

	Before rhEPO therapy	After 6 months of rhEPO therapy
RBC life span, days	20.8 ± 3.4	18.9 ± 5.8

Table 3. Changes of filterability of RBC before and 6 months after rhEPO therapy (n = 6)

	Before rhEPO therapy		After 6 months of rhEPO therapy	
	mean ± SD	%	mean ± SD	%
Filterability index	20.8 ± 6.6	100	24.9 ± 5.4	120.8

Table 4. RBC volume before and 6 months after rhEPO therapy (n = 9)

	Before rhEPO therapy		After 6 months of rhEPO therapy	
	mean ± SD	%	mean ± SD	%
RBC volume, fl	91.7 ± 6.9	100	95.9 ± 4.3	104.6

the exact results of Hct, RBC count, mean corpuscular volume (MCV) and reticulocytes for these data. Looking at table 2, one can see that the RBC life span did not change under rhEPO treatment. With 18.9 ± 5.8 days the RBC life span after 6 months of rhEPO therapy was even lower than before the treatment (20.8 ± 3.4 days). The increase of the filterability index (table 3) from 20.8 ± 6.6 before rhEPO treatment to 24.9 ± 5.4 after 6 months of rhEPO therapy indicates a decreased whole cell deformability. Table 4 shows that the volume of each single RBC increases under the 6 months of rhEPO therapy by about 4.6% compared with the initial value.

Discussion

The data for 6 months after rhEPO therapy for the investigations compared to them before rhEPO treatment was selected in order to be sure all RBC are produced under rhEPO. From the improved membrane properties of the single RBC under rhEPO therapy found in the former study [6] one could expect that the RBC life span improves also. Surprisingly it was not true. RBC life span was even less after 6 months of rhEPO than before. Looking at the standard deviation of the RBC life span after 6 months of rhEPO it can be assumed that the RBC life span remains about the same under rhEPO treatment.

Table 1 shows an increase of the Hct to 123.9% after 6 months of rhEPO treatment. The number of RBC does not increase to this extent (117.7%), which suggests that an increased volume of the RBC must be considered (MCV increased to 105.3%). An unexpected result was the filterability of the RBC after 6 months of rhEPO therapy. The filterability index increased to 120.8% from the initial value. Knowing the results obtained for the single RBC regarding membrane mechanics and deformability [6] it was a contradictory result. Under the supposition of an increased RBC volume the decreased filterability under rhEPO treatment might be explained. Rosenmund et al. [10] have described the decreased RBC filterability in hemodialysis patients related with the shortened erythrocyte half-life. Duchesne-Gueguen et al. [2], Inauen et al. [4] and Udden et al. [13] could not establish relations between filterability and morphological alterations or biochemical changes of the RBC. If the investigation of RBC deformability by measurement of filterability is influenced by the RBC volume, then it has to be discussed whether this method is applicable to assess the real deformability of RBC.

The measurements of the RBC volume by micropipette aspiration confirm the results mentioned above. From table 4 an increase of the RBC volume from 91.7 ± 6.9 fl ($= 100\%$) before rhEPO to 95.9 ± 4.3 fl after 6 months of rhEPO can been seen. The increase of RBC volume to 104.6% corresponds to the findings of MCV (table 1). Reticulocytes of 16.6 ± 3.7 g/l after 6 months of rhEPO increased compared with the initial values (12.4 ± 2.4 g/l) to 133.9% and indicates that the increased RBC volume cannot be referred to a predominating number of young cells [3]. Summarizing these results it is possible to state that the RBC volume increases after 6 months of rhEPO therapy by about 5% (MCV 5.3%, micropipette aspiration 4.6%). Hct increased within the observation period to 123.9% compared to the Hct before rhEPO. This means that about 20% of the increase of Hct under rhEPO therapy results from an increased RBC volume and about 80% from the increased number of RBC.

In conclusion, rhEPO therapy cannot prolong RBC life span. Obviously the uremic milieu alone determines the RBC life time. rhEPO treatment corrects renal anemia by increasing the number of RBC and the RBC volume.

References

1 Akizawa T, Koshikawa S, Takaku F, Urabe A, Akiyama N, Mimura N, Otsubo O, Nihei H, Suzuki Y, Kawaguchi Y, Ota K, Kubo K, Marumo F, Maeda T: Clinical effect of recombinant human erythropoietin on anemia associated with chronic renal failure. A multi-institutional study in Japan. Int J Artif Organs 1988;11:343–350.
2 Duchesne-Gueguen M, Durand F, Le Pogamp P, Garry J, Le Goff M-C, Joyeux V, Chevett D, Genetet B: Study of hemorheological parameters in patients with chronic renal failure and evolution of those parameters in the course of dialysis treatment. Clin Hemorheol 1988;8:407–414.
3 Eschbach JW: Discussion to reference 6, Lerche D, et al.
4 Inauen W, Stäubli M, Descendres C, Gallazzi L, Straub W: Erythrocyte deformability in dialysed and nondialysed uraemic patients. Eur J Clin Invest 1982;12:172–176.
5 Landgraf H, Grützmacher P, Bergmann M, Ehrly AM: Verhalten der Fliesseigenschaften des Blutes bei Patienten mit renaler Anämie im Verlauf einer Erythropoietin-Therapie. Hämostaseologie 1988;8:53–54.
6 Lerche D, Schmidt R, Zoellner K, Meier W, Paulitschke M, Distler B, Klinkmann H: Rheology in whole blood and in red blood cells under recombinant human erythropoietin therapy. Contrib Nephrol. Basel, Karger, 1989, vol 76, pp 299–303.
7 Meier W, Kucera W, Lerche D, Paulitschke M: pH Dependence of human red blood cell elastic membrane properties. Studia Biophys 1985;105:29–38.
8 Meier W, Paulitschke M, Lerche D, Zoellner K, Devaux S: Improvement of mechanical RBC properties assessed with micropipette techniques as a consequence of erythropoietin therapy of juvenile chronically hemodialysed patients. Studia Biophys 1988;128:183–188.
9 Nonnast-Daniel B, Schäffer J, Frei U: Hemodynamics in hemodialysis patients treated with recombinant human erythropoietin. Contrib Nephrol. Basel, Karger, 1989, vol 76, pp 283–289.
10 Rosenmund A, Binswanger U, Straub PW: Oxidative injury to erythrocytes, cell rigidity, and splenic hemolysis in hemodialyzed uremic patients. Ann Intern Med 1975;82:460–465.
11 Schaefer RM, Leschke M, Strauer BE, Heidland A: Blood rheology and hypertension in hemodialysis patients treated with erythropoietin. Am J Nephrol 1988;8:449–453.
12 Steffen HM, Brummer R, Müller R, Degenhardt S, Pollok M, Lang R, Baldamus CA: Peripheral hemodynamics, blood viscosity and the renin-angiotensin system in

hemodialysis patients under therapy with recombinant human erythropoietin. Contrib Nephrol. Basel, Karger, 1989, vol 76, pp 292–298.
13 Udden MM, O'Rear EA, Kegel H, McIntire LV, Lynch EC: Decreased deformability of erythrocytes and increased intracellular calcium in patients with chronic renal failure. Clin Hemorheol 1984;4:473–481.

Doz. Dr. sc. med. Reinhard Schmidt, Department of Internal Medicine,
University of Rostock, Heydemannstrasse 6, D–O–2500 Rostock Germany

rhEPO Treatment of Anemia in Uremic Patients

J. Bommer, H.P. Barth, B. Schwöbel

Medizinische Klinik der Universität, Heidelberg, FRG

The availability of recombinant human erythropoietin (rhEPO) has revolutionized the treatment of anemia in patients with progressed renal failure. The optimal dose and application route are still under discussion. In the following pages, some aspects of renal anemia as well as efficacy, dosage and application route of rhEPO are discussed.

Anemia in Dialysis Patients

Anemia, a typical complication of chronic renal failure, is characterized by a shortened red cell survival and disturbed erythropoiesis. In detail, considerable blood loss contributes to anemia in patients under maintenance hemodialysis. Impaired platelet function, reduced factor VIII activity [1, 2], favors bleeding in the intestinal tract or skin as well as hematoma formation. In addition to the uremic coagulopathy, repetitive heparinization during hemodialysis enhances such bleeding. At the end of hemodialysis, small volumes of blood of about 4 ml remain in tubings, dialyzers and needles. Such small blood loss per dialysis means a total blood loss of 600–700 ml/year in the course of 156 dialysis sessions per year. Blood taken for diagnostic purposes must not be underestimated – 10 ml/week result in 520 ml/year.

Hepatosplenomegaly, more frequent in hemodialysis patients than in healthy controls [3–5], increases erythrocyte sequestration within the spleen. Furthermore, the red cell survival is shortened by mild hemolysis. A cellular defect of erythrocytes seems to play no marked role, since red cell survival is normalized if erythrocytes from a patient with progressed renal failure are infused into a subject with intact renal function [6].

Increased blood loss, low intestinal iron reabsorption and/or insufficient iron supplementation contribute to iron deficiency, present in many non-transfused dialysis patients. Vitamin B_{12} and folic acid deficiency, a very rare finding in our dialysis patients, may also decrease erythropoiesis. Furthermore, transient or persisting diseases other than uremia-like infections, chronic inflammatory status, tumors, etc., can depress erythropoiesis via an inhibitory effect of interleukin-1 and tumor necrosis factor on bone marrow cells.

It has been postulated that uremic inhibitors of erythropoiesis like spermidine [7-9] or acetylspermidine and hypersplenism worsened anemia in patients with renal insufficiency. Aluminium intoxication, a specific complication in patients under maintenance hemodialysis, induces a micro-cytototic anemia [10]. Worsening of anemia has been reported after chloramine, nitrate and copper intoxication in dialysis patients [11, 12]. In 10% of dialysis patients, uremia is the consequence of chronic analgesic abuse. If such abuse is continued during dialysis therapy, anemia can also be worsened. Last, but not least, inappropriate low EPO levels, a common finding in dialysis patients, contribute considerably to renal anemia.

rhEPO Therapy in Uremic Patients

In hemodialysis patients with a hematocrit (Hct) between 20 and 30%, serum EPO levels were comparable with those of healthy controls with a normal Hct of 45-50% [12]. In parallel, the absolute reticulocyte count is also comparable in anemic dialysis patients and healthy controls, suggesting that the endogenous rhEPO is effective in dialysis patients. In experimental studies, Eschbach et al. [14] document a comparable effect of EPO in uremic and nonuremic sheep.

In anemic dialysis patients intravenous rhEPO therapy is followed by a dose-dependent increase of the mean absolute reticulocyte count [15] (fig. 1). Consequently, the mean and median weekly increment of Hct correlates also with the given rhEPO dose. However, in all studies, the response to rhEPO varied considerably from patient to patient. For example, in a German multicenter study, three times 40, 80 or 120 U/kg/week were given to 95 nontransfused hemodialysis patients with a Hct level of <27%. Consequently, a median increase of Hct of 0.64, 1.24 or 1.64% was found after different rhEPO doses (fig. 2). However, at a given rhEPO dose, the increase of Hct varied from 0.2 to 3% per week and the whole range was comparable

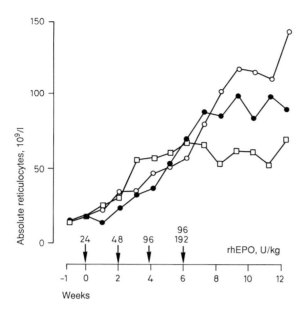

Fig. 1. Mean absolute reticulocyte count in rhEPO-treated dialysis patients. rhEPO therapy was started with 3 × 24 U/kg/week and increased to a final dose of 3 × 48 U/kg/week (□), 3 × 96 U/kg (●) or 192 U/kg (○).

in all groups of patients receiving very different rhEPO doses [16]. In this context, it is of note that serum EPO levels varied by a factor of 10 at a given Hct level in healthy controls [13]. Such a large normal range of serum EPO levels is compatible with the large inter-individual variation of the efficacy of rhEPO response.

In the individual patient, the response to rhEPO cannot be predicted. The response to rhEPO seems to be independent of the basal Hct levels before rhEPO therapy. No significant difference of rhEPO efficacy was found in nontransfused or polytransfused dialysis patients [17]. In patients with a longer dialysis time per week, higher spontaneous Hct levels were reported in a large European survey [18]. It is not yet proved whether the weekly dialysis time modified the efficacy of rhEPO. The rhEPO maintenance dose of 3 × 70 U/kg/week recommended by American authors is higher than the 3 × 35 U/kg/week needed in the German multicenter study or others [19, 20]. Short-time, but highly efficient, dialysis is very common in the USA. Whether this dialysis modality influences the rhEPO maintenance dose remains an open question. The patients in the German study were all

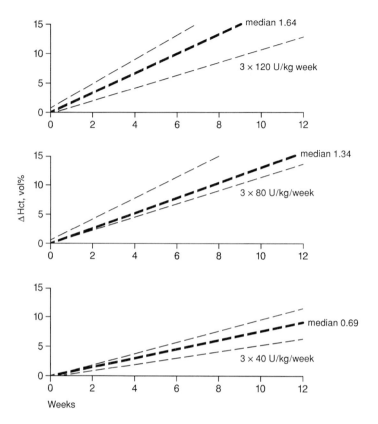

Fig. 2. Increase of Hct under rhEPO therapy if patients were treated constantly with 3 × 40 (80 or 120) U/kg/week (median and interquartile).

dialyzed 3 × 4–5 h/week. Higher Hct levels have been reported in elderly hemodialysis patients compared to younger patients [18]. Any effect of age on the response to rhEPO is not yet known.

A slow increase of Hct levels after rhEPO therapy is of benefit since the hypertensive effect of rhEPO therapy seems to depend on the increase of Hct under such treatment [21]. Even high single rhEPO doses did not induce acute rise of blood pressure. Irrespective of the rhEPO dose, increase of blood pressure was more frequently observed in patients with an abrupt increase of Hct. In a German multicenter study, blood pressure rose only in 23% of patients with an increase of Hct of <0.7% per week, but in 53% of patients with an increase of Hct of >1.5% per week [21].

After 2 years and more of rhEPO therapy the blood pressure and antihypertensive therapy was comparable with that found before EPO therapy was started, despite the fact that the median Hct was increased from 22 to 30% in our long-term studies. This observation suggests that the increase of blood pressure under rhEPO therapy is transient in many patients and the blood pressure is influenced more by the intensity of Hct increase than by the final Hct level under maintenance therapy. Consequently, it is wise to start such therapy with a low rhEPO dose, e.g. 50 U/kg/week i.v., since the response to rhEPO in a single patient cannot be predicted, and a slow increase of Hct is less hazardous.

Application of rhEPO

In several pharmacokinetic studies, a rather low bioavailability of rhEPO was found after subcutaneous (s.c.) compared to intravenous (i.v.) application [22, 23]. After s.c. injection, the mean bioavailability ranged between 21 and 49% of that after i.v. injection in the same patient group. In contrast, we found a surprisingly high efficacy of rhEPO after s.c. therapy in vivo. The weekly rhEPO dose could be reduced markedly if patients under chronic i.v. rhEPO therapy were switched to s.c. for 3–4 months [24]. In a further study we treated 16 other dialysis patients under maintenance i.v. rhEPO therapy later with s.c. rhEPO [25]. As shown in figure 3, Hct levels were fairly constant under i.v. or s.c. rhEPO therapy, despite a marked dose reduction of rhEPO during the s.c. treatment period (fig. 4). If rhEPO was given about twice a week, i.v. or s.c., the required s.c. dose was about 50–60% of the i.v. dose. A comparable reticulocyte count under i.v. or s.c. rhEPO therapy confirms the high efficacy of rhEPO given s.c.

Similar to insulin, EPO was injected s.c. in the upper or lower arm, upper leg or abdominal area. Subcutaneous rhEPO therapy was well tolerated in dialysis patients. Several patients complained about the additional sticks after dialysis. Only rare and small hematomas were observed if the drug was given s.c. at the end of hemodialysis to a heparinized patient.

In 1 patient with a history of atopy including severe allergic reactions to ethylene oxide and others, we observed intensive urticaria and local rubor at the injection sites after s.c. application of rhEPO. This patient tolerated the i.v. EPO therapy without any complications over >2 years and no EPO antibodies could be detected.

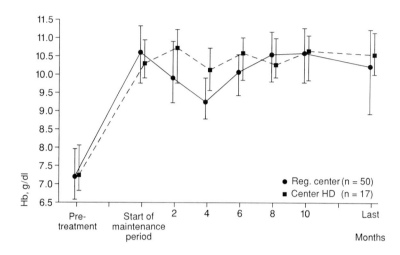

Fig. 3. Hemoglobin under i.v. or s.c. rhEPO therapy (mean and standard deviation).

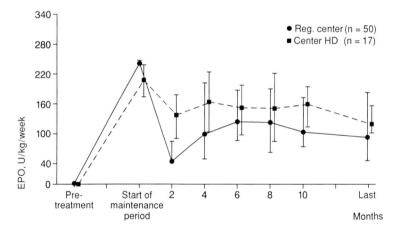

Fig. 4. Maintenance dose of rhEPO given i.v. or s.c. to obtain Hct levels shown in figure 3 (mean and standard deviation).

Such a high efficacy of s.c. rhEPO agrees with several other studies. Stevens et al. [19] reported a maintenance s.c. rhEPO dose of 3 times 25 U/kg compared with 3 times 48 U/kg i.v. in different groups of dialysis patients. In preuremic patients, Eschbach et al. [26] reported a comparable increase of Hct under 3 times 100 U rhEPO s.c. and 150 U i.v. In a limited number of

Table 1. Cost-saving effect of s.c. rhEPO therapy

Total dialysis patients (FRG)	n = 20,000	
rhEPO-treated	n = 6,000	
rhEPO dose (i.v.)	100 U/kg/week	150 U/kg/week
Costs Per patient/year (i.v.)	14,800 DM	22,200 DM
Total FRG/year	88.8 mio. DM	133.2 mio. DM
Costs saved (s.c.)	31 mio. DM	46.6 mio. DM

patients, rhEPO maintenance dose could be reduced by about 60% if the drug was given by daily s.c. injection [27]. Such a further reduction of rhEPO dose under daily s.c. injection must be balanced against mild discomfort of daily injection and wastage of rhEPO preparing the daily dose if a new vial must be opened every day. Furthermore, the patient must be able to inject the EPO himself.

It is rather difficult to reconcile the pharmacokinetic data and clinical effect of s.c. rhEPO. The low bioavailability calculated as the area under the curve after s.c. rhEPO may be influenced by a delayed s.c. resorption and concomitant fast absorption of rhEPO in the bone marrow cell. The increased erythropoiesis in the presence of mildly elevated EPO serum levels under s.c. rhEPO treatment agrees with recent reports on rhEPO levels and improvement of anemia after kidney transplantation. Except for one or two transient rises, serum EPO levels were only mildly elevated, but hematocrit continues to increase over several weeks in patients with functioning renal graft.

In general, the maintenance dose of s.c. rhEPO was 30–50% less than the i.v. dose. Such a dose reduction has a marked cost-saving effect (table 1). In addition to the cost-saving effect of such s.c. rhEPO, application seems to be more physiological. Intravenous application of rhEPO is followed by extremely high peak levels not occurring if an identical rhEPO dose is given s.c. In contrast, a moderate persisting elevation of rhEPO over several days is found after s.c. application. In patients after kidney transplantation, Hct levels were doubled after several weeks in the presence of only mildly increased rhEPO serum levels, in good agreement with the stimulated erythropoiesis following a borderline, but persisting elevation of serum EPO levels under s.c. rhEPO therapy.

If EPO has any proliferative effect on any cells other than erythroblasts, perhaps even malignant cells, extremely high peak levels after i.v. application may be of more concern than moderate serum levels after s.c. injection.

References

1 Andrassy K, Ritz E, Bommer J: Effects of hemodialysis on platelets. Contrib Nephrol. Basel, Karger, 1987, vol 59, pp 26–34.
2 Janson PA, Jubiliere SJ, Weinstein MJ, Deykin D: Treatment of the bleeding tendency in uremia with cryoprecipitate. New Engl J Med 1980;303:1318–1322.
3 Bommer J, Ritz E, Waldherr R: Silicone-induced splenomegaly. Treatment of pancytopenia by splenectomy in a patient on hemodialysis. New Engl J Med 1981; 305:1077–1079.
4 Anastasiades S, Sheriff M, Platts D, Bartolo D: Splenomegaly in patients receiving regular hemodialysis. Kidney Int 1981;19:619A.
5 Bommer J, Waldherr R, Gastner M, Lemmes R, Ritz E: Iatrogenic multiorgan silicone inclusions in dialysis patients. Klin Wochenschr 1981;59:1149–1157.
6 Chaplin H, Mollison PL: Red cell life span in nephritis and in hepatic cirrhosis. Clin Sci 1953;12:351–360.
7 McGonigle RJS, Husserl F, Wallin JD, Fischer JW: Hemodialysis and continuous ambulatory peritoneal dialysis effects on erythropoiesis in renal failure. Kidney Int 1984;25:430–436.
8 Radtke HW, Rege AW, LaMarche MB, Bartos D, Campbell RA, Fisher JW: Identification of spermine as an inhibitor of erythropoiesis in patients with chronic renal failure. J Clin Invest 1980;67:1623–1629.
9 Segal GM, Stueve T, Adamson JW: Spermine and spermidine are non-specific inhibitors of in vitro hematopoiesis. Kidney Int 1987;31:72–76.
10 Short AIK, Winney RJ, Robson JS: Reversible microcytic hypochromic anaemia in dialysis patients due to aluminium intoxication. Proc EDTA 1980;17:226–232.
11 Bommer J, Ritz E: Water quality – a neglected problem in hemodialysis. Nephron 1987;46:1–6.
12 Lyle WH, Payton JE, Hui M: Haemodialysis and copper fever. Lancet 1976;i:1324–1325.
13 Erslev AJ, Caro J: Erythropoietin titers in response to anemia or hypoxia. Blood Cells 1987;13:207–216.
14 Eschbach J, Mladenovic J, Garcia JF, Wahl PW, Adamson JW: The anemia of chronic renal failure in sheep: response to erythropoietin-rich plasma in vivo. J Clin Invest 1984;74:434–441.
15 Bommer J, Alexiou C, Müller-Bühl U, Eifert J, Ritz E: Recombinant human erythropoietin therapy in haemodialysis patients – dose determination and clinical experience. Nephrol Dial Transplant 1987;2:238–242.
16 Pollok M, Bommer J, Gurland HJ, Koch KM, Schoeppe W, Scigalla P, Baldamus CA: Effects of recombinant human erythropoietin treatment in end-stage renal failure patients. Contrib Nephrol. Basel, Karger, 1989, vol 76, pp 201–211.
17 Cilag European Multicenter Study on rhEPO Therapy in Dialysis Patients, 1986–88.

18 EDTA Report, 1989.
19 Stevens C, Strang A, Oliver WO, Winearls CG, Cotes PM: Subcutaneous erythropoietin and peritoneal dialysis. Lancet 1989;i:1388–1389.
20 Lui SF, Chung WWM, Leung CB, Chan K, Lai KN: Pharmacokinetics and pharmacodynamics of subcutaneous and intraperitoneal administration of recombinant human erythropoietin in patients on continuous ambulatory peritoneal dialysis. Clin Nephrol 1990;33:47–51.
21 Samtleben W, Baldamus CA, Bommer J, Fassbinder W, Nonnast-Daniel B, Gurland HJ: Blood pressure changes during treatment with recombinant human erythropoietin. Contrib Nephrol. Basel, Karger, 1988, vol 66, pp 114–122.
22 MacDougall I, Roberts DE, Neubert P, Dharmasena AD, Coles GA, Williams JD: Pharmacokinetics of recombinant human erythropoietin in patients on continuous ambulatory peritoneal dialysis. Lancet 1989;i:425–427.
23 Kampf D, Kahl A, Passlick J, Pustelnik A, Eckhardt KU, Ehmer B, Jacobs C, Baumelou A, Grabensee B, Gahl GM: Single-dose kinetics of recombinant human erythropoietin after intravenous, subcutaneous and intraperitoneal administration. Contrib Nephrol. Basel, Karger, 1989, vol 76, pp 106–111.
24 Bommer J, Müller-Bühl E, Ritz E, Eifert J: Recombinant human erythropoietin in anaemic patients on haemodialysis. Lancet 1987;i:392.
25 Bommer J, Samtleben W, Koch KM, Baldamus CA, Grützmacher P, Scigalla P: Variations of recombinant human erythropoietin application in hemodialysis patients. Contrib Nephrol. Basel, Karger, 1989, vol 76, pp 149–158.
26 Eschbach JW, Kelly MR, Haley NR, Abels RI, Adamson JW: Treatment of the anemia of progressed renal failure with recombinant human erythropoietin. New Engl J Med 1989;321:158–163.
27 Granolleras C, Branger B, Beau MC, Deschodt G, Alsabadani B, Shaldon S: Experience with daily self-administered subcutaneous erythropoietin. Contrib Nephrol. Basel, Karger, 1989, vol 76, pp 143–148.

Prof. Dr. med. J. Bommer, Medizinische Klinik der Universität,
Bergheimer Strasse 58, D-6900 Heidelberg (FRG)

Modulation of the Production of Erythropoietin by Cytokines: In vitro Studies and Their Clinical Implications[1]

Wolfgang Jelkmann[a], *Martin Wolff*[a], *Joachim Fandrey*[b]

[a] Physiologisches Institut I, Rheinische Friedrich-Wilhelms-Universität, Bonn;
[b] Physiologisches Institut, Medizinische Universität, Lübeck, BRD

The production of erythropoietin (EPO) is greatly stimulated when the whole body O_2 offer is lowered. In humans subjected to hypoxia the plasma EPO level increases in dependence of the degree of the lowering of the inspiratory O_2 tension [1]. In anemic humans without renal disease, an inverse relationship exists essentially between the plasma EPO level and the hematocrit [2] or the hemoglobin concentration of the blood [3]. However, clinical studies have provided some evidence which suggests that the level of EPO is relatively low for the degree of the anemia in many patients suffering from chronic inflammation or malignancy [4–9]. Noteworthy, too, is that in chronic renal failure hematocrit values vary considerably among different patients with similar glomerular filtration rates. Hence it has been proposed that as yet undefined factors may influence the production of EPO independently of changes in the O_2 offer [10].

Our working hypothesis is that the immune system plays a modulating role in the control of the production of EPO. Immunological reactions and inflammatory processes are associated with increased local production and release of hormone-like polypeptides, cytokines, by activated macrophages and lymphocytes.

Herein, we first describe the effects of the immunomodulatory peptides interleukin-1 (IL-1), tumor necrosis factor-α (TNF-α) and interferon-γ (IFN-γ) on the production of EPO in hepatoma cell cultures (HepG2). Hepatoma cells were used because there are still no renal cell cultures available that are capable of controlled EPO gene expression. Second, we

---

[1] Supported in part by DFG Grant Je 95/6-2.

have made an attempt to summarize the clinical observations of anomalously low or high EPO levels that may be related to immune reactions.

In vitro Studies

Materials and Methods

The following human recombinant immunomodulatory peptides were tested for their effects on the in vitro production of EPO: IL-1α, IL-1β, TNF-α and IFN-γ (all from Boehringer Mannheim, FRG). The studies were carried out with hepatoma cells of the line HepG2, which has the capacity for controlled production of EPO [11]. The cells were obtained from the American Type Culture Collection (ATCC No. HB 8065). They were repeatedly subcultured and grown in medium RPMI 1640 (Boehringer), supplemented with 10% fetal bovine serum (Gibco BRL, Eggenstein, FRG) and sodium bicarbonate (2.2 g/l). With the exception noted below, the cultures were plated in 24-well polystyrol dishes (2 cm^2 bottom; Falcon, Becton-Dickinson, Heidelberg, FRG) and incubated at 37 °C in humidified air with 5% CO_2 (Heraeus Incubators, Hanau, FRG). Because the polystyrol bottom of these dishes is almost impermeable for O_2, the cells produce EPO here under conditions of diffusion-limited O_2 supply. In order to demonstrate the stimulation of the production of EPO by low (2% O_2) versus high (21% O_2) concentration of O_2 in the incubation atmosphere, additional studies were carried out using cultures grown in specific 20-cm^2 dishes with a gas-permeable bottom (Petriperm hydrophilic; Heraeus). All experiments were performed with confluent monolayers (about 0.5×10^6 cells/cm^2).

The experiments were started by the addition of fresh medium (0.5 ml/cm^2) and the respective cytokine. Twenty-four hours thereafter, the culture medium was collected for the assay of EPO. The cell layers were washed with saline and then lysed with SDS-NaOH (sodium dodecyl sulfate, 5 g/l, in NaOH, 0.1 mol/l) for determination of the amount of cellular protein (Micro protein determination kit; Sigma Diagnostics, Taufkirchen, FRG).

EPO was measured by radioimmunoassay using ^{125}I-labeled recombinant human EPO (rhuEPO, 11-33 TBq/mmol; Amersham Buchler, Braunschweig, FRG), anti-serum from a rabbit previously immunized with rhuEPO (Cilag, Sulzbach, FRG) and human urinary EPO (kindly provided by the National Institutes of Health, Md., USA) as the standard. Following 24 h of incubation, the free and antibody-bound ^{125}I-rhuEPO were separated by precipitation with polyethylene glycol. The detection limit of the assay was 5 mU/ml, the intraassay variance 5.2% and the interassay variance 12.0% at 20 mU/ml.

In addition, in distinct cultures the rate of the production of α-fetoprotein was determined in order to test the EPO specificity of the observed effects. α-Fetoprotein was measured using a commercial ^{125}I-labeled radioimmunoassay kit (Pharmacia, Uppsala, Sweden).

The data are given as the mean ± standard deviation. Dunnett's test was applied to determine the significance of difference ($p < 0.05$) between a control mean and several treatment means.

Results

The pO_2 dependence of the synthesis of EPO by the present HepG2 cells was assured in cultures grown in dishes with a gas-permeable bottom. Here,

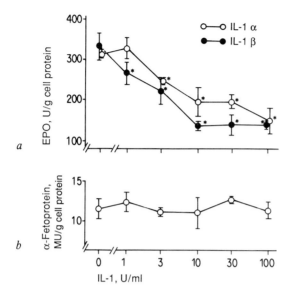

Fig. 1. Effects of the addition of IL-1α or IL-1β on the 24-hour rates of the production of EPO *(a)* and α-fetoprotein *(b)* in HepG2 cultures. Mean ± SD of 4 parallel cultures. *Significantly different from control.

significantly more EPO was produced when the O_2 concentration in the incubation atmosphere was reduced from 21% to 2% (12.9 ± 1.7 compared to 40.2 ± 2.8 U EPO/g cell protein in 24 h; n = 4).

Figure 1 shows that the addition of IL-1 resulted in a dose-dependent inhibition of the production of EPO in HepG2 cultures (diffusion-limited O_2 supply). IL-1β was effective at lower concentrations than IL-1α. The production of α-fetoprotein was not affected by IL-1 (only IL-1α was tested).

Figure 2 shows that TNF-α also reduced the production of EPO, but not of α-fetoprotein (diffusion-limited O_2 supply).

Figure 3 shows that IFN-γ, in contrast to IL-1 and TNF-α, tended to increase the formation of EPO in HepG2 cultures (diffusion-limited O_2 supply). The production of α-fetoprotein was slightly reduced with high doses of IFN-γ.

Diseases Associated with Abnormal Blood Levels of EPO

The kidney is the primary site of the production of EPO in adults [reviewed in 12]. In addition, EPO mRNA is expressed in the liver during

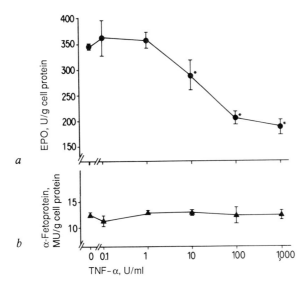

Fig. 2. Effects of the addition of TNF-α on the 24-hour rates of the production of EPO *(a)* and α-fetoprotein *(b)* in HepG2 cultures. Mean ± SD of 4 parallel cultures. *Significantly different from control.

severe hypoxia [13]. Anemic patients with chronic renal failure may have higher than normal plasma EPO levels [10, 14, 15], and a further increase occurs upon exposure to acute hypoxic stress [10]. Nevertheless, in end-stage renal disease the plasma EPO level is inappropriately low in relation to the blood hemoglobin concentration [10, 14–16]. It is of interest that the plasma EPO concentration may increase after switching from hemodialysis to continuous ambulatory peritoneal dialysis in chronic renal failure [17].

In anemias not associated with renal insufficiency, a fairly strong negative correlation exists between the EPO and the hemoglobin concentration in the blood [3]. It was stated earlier that the production of EPO in anemic subjects with normal kidney function depends only on the O_2 carrying capacity of the blood and not on the specific disorder causing the anemia [2].

However, several exceptions to this rule have also been reported. Some examples are shown in table 1. Levels of EPO lower than expected for the severity of anemia were measured in most studies of patients with malignancies [4, 6, 8, 9, 18, 19], e.g. in lung cancer [8, 9], and in experimental animals bearing extramedullary solid tumors [20]. However, in most of these studies little distinction was made between different types and stages of tumor

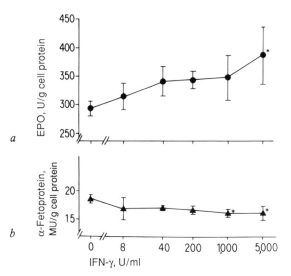

Fig. 3. Effects of the addition of IFN-γ on the 24-hour rates of the production of EPO *(a)* and α-fetoprotein *(b)* in HepG2 cultures. Mean ± SD of 4 parallel cultures. *Significantly different from control.

Table 1. Diseases in which lowered or increased blood EPO levels have been reported

Change in EPO level	Disease	References
Decrease	chronic renal failure	10, 14–16
	acute renal graft rejection	29
	malignancy	4, 6, 8, 9, 18, 19
	chronic infection,	4–6
	chronic inflammation	
	(e.g. rheumatoid arthritis)	5–7, 22–25
	AIDS	28
Increase	hepatitis in renal failure	32, 33
	aplastic anemia,	30
	leukemia with erythrocytic	
	hypoplasia	31

disease, and the EPO values varied greatly among the individual patients. It is not surprising, therefore, that some investigators estimated normally raised EPO levels in malignant diseases [5], including lung cancer [21]. Relatively low levels of bioactive EPO were generally observed in patients with anemias secondary to chronic bacterial infection [4–6] or chronic

Table 2. Immune reactions and some specific disorders in which increased IL-1 and TNF-α production have been reported

Cytokine	Immune reactions	Specific diseases	References
IL-1	antigen-antibody reaction, inflammation, fever	acute infection rheumatoid arthritis glomerulonephritis	reviewed in 34 36 37
TNF-α	inflammation, fever, cachexia, tissue remodelling	acute and chronic infection rheumatoid arthritis acute renal graft rejection malignancy AIDS	reviewed in 35 38 39 40 41

inflammation [5–7, 22, 23]. Because most of the earlier reports included a miscellany of diseases, it is important to note that recent studies utilizing sensitive radioimmunoassays of EPO confirmed these observations in patients with rheumatoid arthritis [24, 25] with one exception [26]. Moreover, the replacement therapy with rhuEPO has proved useful in the treatment of the anemia of rheumatoid arthritis [27]. In addition, of interest are the inappropriately low serum EPO levels observed in HIV-infected patients, because once their infection is treated with Zidovudine, their EPO levels increase greatly in response to the anemia [28]. Another example suggesting a link between immune reactions and the production of EPO is the fall in the serum EPO level during episodes of acute allograft rejection in renal transplant recipients [29].

Vice versa, some diseases are associated with increased EPO production. Here, the extremely high serum EPO levels are observed in patients with aplastic anemia [30] or leukemia with marrow hypoplasia [31]. Finally, the spontaneous increase in serum EPO is of interest which follows viral or toxic hepatitis in uremic patients undergoing hemodialysis [32, 33].

Immune responses are associated with the production of immunomodulatory peptides by leukocytes and tissue macrophages which act as local and systemic mediators in the pathogenesis of infection, allergies, inflammation and cachexia. Among the best studied cytokines are IL-1 [34] and TNF-α [35]. A random look at the literature reveals that these immunomodulatory peptides play an important role in the pathogenesis of several diseases that are characterized by anomalously low EPO levels (compare table 1 and table 2).

IL-1 and TNF-α have been regarded as mediators of the anemia of chronic disorders [42–44]. Treatment of mice [44], rats [43] and humans in phase I studies [45] with TNF-α and treatment of mice with IL-1 [46] resulted in lowered red cell production. It remains to be clarified whether these effects are only due to inhibition of the action of EPO in erythropoietic tissues or if the synthesis of EPO is also inhibited. To the best of our knowledge, there are only two reports on the effects of cytokines on in vivo EPO production. Human TNF-α failed to lower the blood EPO level in anemic mice [44]. Murine IFN-α attenuated the EPO response to hypobaric hypoxia in mice [47].

Summary and Conclusions

The etiology of the anemia of chronic disorders is complex. Factors which clearly contribute to the suppression of erythropoiesis are (a) reduced iron availability and (b) stimulation of the synthesis of immunomodulatory peptides such as IL-1, TNF-α and IFN-γ, which inhibit the proliferation of erythrocytic progenitors in the bone marrow [42, 44, 46, 48, 49]. The question as to whether lack of EPO is of general importance in the pathogenesis of the anemia of chronic inflammatory and malignant diseases is still a subject of controversy [50].

The present in vitro studies show that IL-1 and TNF-α, but not IFN-γ significantly lower the pO_2-dependent formation of EPO in HepG2 cultures. In addition, clinical examples are given of anomalously low or high EPO levels in association with diseases involving the immune system. It is proposed that monokines and related immunomodulatory peptides could play a role in the control of the production of EPO.

References

1 Eckhardt KU, Boutellier U, Kurtz A, Schopen M, Koller EA, Bauer C: Rate of erythropoietin formation in humans in response to acute hypobaric hypoxia. J Appl Physiol 1989;66:1785–1788.
2 Erslev AJ, Wilson J, Caro J: Erythropoietin titers in anemic, nonuremic patients. J Lab Clin Med 1987;109:429–433.
3 Jelkmann W, Wiedemann G: Serum erythropoietin level: relationships to blood hemoglobin concentration and erythrocytic activity of the bone marrow. Klin Wochenschr 1990;68:403–407.

4 Ward HP, Kurnick JE, Pisarczyk MJ: Serum level of erythropoietin in anemias associated with chronic infection, malignancy, and primary hematopoietic disease. J Clin Invest 1971;50:332–335.
5 Zucker S, Friedman S, Lysik RM: Bone marrow erythropoiesis in the anemia of infection, inflammation, and malignancy. J Clin Invest 1974;53:1132–1138.
6 Douglas SW, Adamson JW: The anemia of chronic disorders: studies of marrow regulation and iron metabolism. Blood 1975;45:55–65.
7 Pavlović-Kentera V, Ruvidić R, Milenković P, Marinković D: Erythropoietin in patients with anaemia in rheumatoid arthritis. Scand J Hematol 1979;23:141–145.
8 Dainiak N, Kulkarni V, Howard D, Kalmanti M, Dewey MC, Hoffman R: Mechanisms of abnormal erythropoiesis in malignancy. Cancer 1983;51:1101–1106.
9 Cox R, Musial T, Gyde OH: Reduced erythropoietin levels as a cause of anaemia in patients with lung cancer. Eur J Cancer Clin Oncol 1986;22:511–514.
10 Chandra M, Clemons GK, McVicar MI: Relation of serum erythropoietin levels to renal excretory function: evidence for lowered set point for erythropoietin production in chronic renal failure. J Pediatr 1988;113:1015–1021.
11 Goldberg MA, Glass GA, Cunningham JM, Bunn HF: The regulated expression of erythropoietin by two human hepatoma cell lines. Proc Natl Acad Sci USA 1987;84: 7972–7976.
12 Jelkmann W: Renal erythropoietin: properties and production. Rev Physiol Biochem Pharmacol 1986;104:139–215.
13 Bondurant MC, Koury MJ: Anemia induces accumulation of erythropoietin mRNA in the kidney and liver. Mol Cell Biol 1986;6:2731–2733.
14 Caro J, Brown S, Miller O, Murray T, Erslev AJ: Erythropoietin levels in uremic nephric and anephric patients. J Lab Clin Med 1979;93:449–458.
15 Radtke HW, Claussner A, Erbes PM, Scheuermann EH, Schoeppe W, Koch KM: Serum erythropoietin concentration in chronic renal failure: relationship to degree of anemia and excretory renal function. Blood 1979;54:877–884.
16 Cotes PM, Pippard MJ, Reid CDL, Winearls CG, Oliver DO, Royston JP: Characterization of the anaemia of chronic renal failure and the mode of its correction by a preparation of human erythropoietin (r-HuEPO). An investigation of the pharmacokinetics of intravenous erythropoietin and its effects on erythrokinetics. Q J Med 1989;70:113–137.
17 Chandra M, McVicar M, Clemons G, Mossey RT, Wilkes BM: Role of erythropoietin in the reversal of anemia of renal failure with continuous ambulatory peritoneal dialysis. Nephron 1987;46:312–315.
18 Firat D, Banzon J: Erythropoietic effect of plasma from patients with advanced cancer. Cancer Res 1971;31:1353–1359.
19 Schwartz ML, Israel HL: Severe anemia as a manifestation of metastatic jugular paraganglioma. Arch Otolaryngol 1983;109:269–272.
20 Degowin RL, Gibson DP: Erythropoietin and the anemia of mice bearing extramedullary tumor. J Lab Clin Med 1979;94:303–311.
21 Schreuder WO, Ting WC, Smith S, Jacobs A: Testosterone, erythropoietin and anaemia in patients with disseminated bronchial cancer. Br J Haematol 1984;57: 521–526.
22 Ward HP, Gordon B, Pickett JC: Serum levels of erythropoietin in rheumatoid arthritis. J Lab Clin Med 1969;74:93–97.
23 Vichinsky EP, Pennathur-Das R, Nickerson B, Minor M, Kleman K, Higashino S, Lubin B: Inadequate erythroid response to hypoxia in cystic fibrosis. J Pediatr 1984; 105:15–21.

24 Baer AN, Dessypris EN, Goldwasser E, Krantz SB: Blunted erythropoietin response to anaemia in rheumatoid arthritis. Br J Haematol 1987;66:559–564.
25 Hochberg MC, Arnold CM, Hogans BB, Spivak JL: Serum immunoreactive erythropoietin in rheumatoid arthritis: impaired response to anemia. Arthritis Rheum 1988; 31:1318–1321.
26 Birgegård G, Hällgren R, Caro J: Serum erythropoietin in rheumatoid arthritis and other inflammatory arthritides: relationship to anaemia and the effect of anti-inflammatory treatment. Br J Haematol 1987;65:479–483.
27 Means RT, Olsen NJ, Krantz SB, Dessypris EN, Graber SE, Stone WJ, O'Neil VL, Pincus T: Treatment of the anemia of rheumatoid arthritis with recombinant human erythropoietin: clinical and in vitro studies. Arthritis Rheum 1989;32:638–642.
28 Spivak JL, Barnes DC, Fuchs E, Quinn TC: Serum immunoreactive erythropoietin in HIV-infected patients. JAMA 1989;261:3104–3107.
29 Rejman ASM, Grimes AJ, Cotes PM, Mansell MA, Joekes AM: Correction of anaemia following renal transplantation: serial changes in serum immunoreactive erythropoietin, absolute reticulocyte count and red-cell creatine levels. Br J Haematol 1985;61:421–431.
30 Pavlović-Kentera V, Milenković P, Ruvidić R, Jovanović V, Biljanović-Paunović L: Erythropoietin in aplastic anemia. Blut 1979;39:345–350.
31 Johannsen H, Jelkmann W, Wiedemann G, Otte M, Wagner T: Erythropoietin/haemoglobin relationship in leukaemia and ulcerative colitis. Eur J Haematol 1989; 43:201–206.
32 Brown S, Caro J, Erslev AJ, Murray TC: Spontaneous increase in erythropoietin and hematocrit value associated with transient liver enzyme abnormalities in an anephric patient undergoing hemodialysis. Am J Med 1980;68:280–284.
33 Simon P, Meyrier A, Tanquerel T, Ang K-S: Improvement of anaemia in haemodialysed patients after viral or toxic haptic cytolysis. Br J Med 1980;280:892–894.
34 Duff GW: Peptide regulatory factors in non-malignant disease. Lancet 1989;i:1432–1435.
35 Tracey KJ, Vlassara H, Cerami A: Cachectin/tumour necrosis factor. Lancet 1989;i: 1122–1126.
36 Eastgate JA, Symons JA, Wood NC, Grinlinton FM, Di Giovine FS, Duff GW: Correlation of plasma interleukin-1 levels with disease activity in rheumatoid arthritis. Lancet 1988;ii:706–709.
37 Matsumoto K, Dowling J, Atkins RC: Production of interleukin-1 in glomerular cell cultures from patients with rapidly progressive crescentic glomerulonephritis. Am J Nephrol 1988;8:463–470.
38 Saxne T, Palladino A, Heinegard D, Talal N, Wollheim A: Detection of tumor necrosis factor-α but not tumor necrosis factor-β in rheumatoid arthritis synovial fluid and serum. Arthritis Rheum 1988;31:1041–1045.
39 Maury CPJ, Teppo AM: Raised serum levels of cachectin/tumor necrosis factor-α in renal allograft rejection. J Exp Med 1987;166:1132–1137.
40 Balkwill F, Osborne R, Burke F, Naylor S, Talbot D, Durbin H, Tavernier J, Fiers W: Evidence for tumour necrosis factor/cachectin production in cancer. Lancet 1987;ii: 1229–1232.
41 Lahdevirta J, Maury CPJ, Teppo A-M, Repo H: Elevated levels of circulating cachectin-tumor necrosis factor in patients with acquired immunodeficiency syndrome. Am J Med 1988;85:289–291.

42 Schooley JC, Kullgren B, Allison AC: Inhibition by interleukin-1 of the action of erythropoietin on erythroid precursors and its possible role in the pathogenesis of hypoplastic anaemias. Br J Haematol 1987;67:11–17.

43 Tracey KJ, Wei H, Manogue KR, Fong Y, Hesse DG, Nguyen HT, Kuo GC, Beutler B, Cotran RS, Cerami A, Lowry SF: Cachectin/tumor necrosis factor induces cachexia, anemia, and inflammation. J Exp Med 1988;167:1211–1227.

44 Johnson RA, Waddelow TA, Caro J, Oliff A, Roodman GD: Chronic exposure to tumor necrosis factor in vivo preferentially inhibits erythropoiesis in nude mice. Blood 1989;74:130–138.

45 Blick M, Sherwin SA, Rosenblum M, Gutterman J: Phase I study of recombinant tumor necrosis factor in cancer patients. Cancer Res 1987;47:2986–2989.

46 Johnson CS, Keckler DJ, Topper MI, Braunschweiger PG, Furmanski P: In vivo hematopoietic effects of recombinant interleukin-1α in mice: stimulation of granulocytic, monocytic, megakaryocytic, and early erythroid progenitors, suppression of late-stage erythropoiesis, and reversal of erythroid suppression with erythropoietin. Blood 1989;73:678–683.

47 Huie ML, Gordon AS, Mirand EA, Leong S, Preti RA, Naughton BA: The effects of interferon on murine erythropoiesis. Life Sci 1985;36:2459–2462.

48 Johnson CS, Chang M-J, Furmanski P: In vivo hematopoietic effects of tumor necrosis factor-α in normal and erythroleukemic mice: characterization and therapeutic applications. Blood 1988;72:1875–1883.

49 Mamus SW, Beck-Schroeder S, Zanjani ED: Suppression of normal human erythropoiesis by gamma-interferon in vitro. J Clin Invest 1985;75:1496–1503.

50 Birgegård G: Erythropoiesis and inflammation. Contrib Nephrol. Basel, Karger, 1989, vol 76, pp 330–341.

Prof. Dr. med. Wolfgang Jelkmann, Physiologisches Institut I,
Rheinische Friedrich-Wilhelms-Universität, Nussallee 11, D–5300 Bonn (FRG)

Iron Status of Dialysis Patients under rhuEPO Therapy

Walter H. Hörl, Karl Dreyling, Hjalmar B. Steinhauer, Rupert Engelhardt, Peter Schollmeyer

Department of Medicine, University of Freiburg i.Br., FRG

Recently, Eschbach et al. [1] summarized a phase III clinical multicenter trial of recombinant human erythropoietin (rhuEPO) therapy undertaken in the United States, where 333 hemodialysis patients entered the study. There were two important clinical observations with respect to iron status: iron deficiency and reduction of iron overload.

Iron Deficiency

Of these 333 patients, 142 (43%) developed evidence of absolute or relative iron deficiency [1]. The term 'relative iron deficiency' designates an iron supply that would be adequate to meet the needs of basal erythropoiesis but which is not adequate for the needs of an expanded erythroid marrow. Iron deficiency occurred at any time during rhuEPO therapy, depending on the quantity of iron stores at the beginning of therapy [1]. Iron deficiency was defined (a) by a serum ferritin of <50 ng/ml and a percent transferrin saturation <20 or (b) by a percent transferrin saturation <20 and normal serum ferritin.

Serum iron declined during rhuEPO therapy in several studies [2–8]. Serum ferritin values also decreased in nearly every patient during rhuEPO therapy [2, 3, 7, 9–13]. In single cases an increase of serum ferritin was observed due to infection, liver disease, chronic inflammation, malignancy, or redistribution from the liver to plasma. Polytransfused patients respond better with their rise in hematocrit values when rhuEPO therapy is instituted than patients who had not required transfusions. This seems to be due to the

Table 1. Serum, plasma and intracellular iron parameters of chronically uremic hemodialysis patients before and after 3 months of rhuEPO therapy

Parameter	Before rhuEPO	After rhuEPO
Serum ferritin, µg/l	1,145 ± 496	251 ± 52*
Erythrocyte ferritin, µg/l	232 ± 127	105 ± 40*
Plasma lactoferrin, µg/l	192 ± 29	165 ± 24
Neutrophil lactoferrin, µg/l	14.3 ± 4.8	8.0 ± 1.3

Mean values ± SEM from 12 patients. *$p < 0.05$.

fact that transfusion-dependent patients had higher iron stores than those not receiving blood transfusions [14]. Therefore, lower rhuEPO doses are recommended for iron overloaded, chronically uremic patients. The present study was performed on hemodialysis patients undergoing rhuEPO treatment with respect to serum, plasma and intracellular iron parameters.

Patients and Methods

Twelve chronically uremic patients with a mean age of 43.1 ± 3.7 years undergoing regular hemodialysis treatment for 69.2 ± 11.7 months (range 4–143) were studied. Hemodialysis was performed 3 times weekly for 4.1 ± 0.2 h using dialyzers made of polymethylmethacrylate (Toray, Tokyo, Japan). The patients received rhuEPO intravenously 3 times weekly in a mean dose of 50 U/kg body weight (b.w.) at the beginning and of 68.3 ± 8.9 U/kg b.w. after 3 months. rhuEPO caused an increase of erythrocytes from 2.5 ± 0.1 to 3.6 ± 0.1 million cells/µl, of hemoglobin from 7.5 ± 0.2 to 10.8 ± 0.3 g/dl, and of hematocrit from 22.5 ± 0.6 to 33.0 ± 1.0%, respectively.

Determination of serum ferritin was carried out using an immunoradiometric assay (RIA, PRIST; Pharmacia, Sweden) with ^{125}I-labelled antiferritin antibodies. Erythrocyte ferritin was measured with the ELISA technique as described by Bodemann et al. [15]. Plasma and neutrophil lactoferrin values were determined with an enzyme-linked immunosorbent assay as described by Rautenberg et al. [16]. PMNs were prepared from 10 ml whole blood anticoagulated with 1.5% Na-EDTA in phosphate-buffered saline (PBS) as described by Harbeck et al. [17].

Results

Serum and intracellular iron parameters of hemodialysis patients before and after 3 months of rhuEPO therapy are shown in table 1. There is a

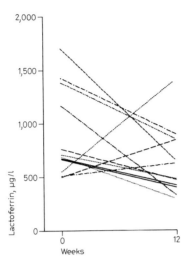

Fig. 1. Lactoferrin at the end of hemodialysis before and 12 weeks after rhuEPO treatment (n = 12).

decrease of serum ferritin, erythrocyte ferritin and neutrophil lactoferrin, whereas plasma lactoferrin was unchanged.

Figure 1 shows individual lactoferrin plasma values at the end of hemodialysis treatment before and 12 weeks after rhuEPO treatment. Plasma lactoferrin values decreased in 9 out of 12 patients and increased in the other 3 subjects.

We also studied 14 patients with end-stage renal failure who had been on CAPD for 25.4 ± 2.9 months. Serum ferritin of these patients during 1 year of rhuEPO therapy is shown in table 2. Before rhuEPO treatment, 6 out of 14 patients were orally supplemented with 40–80 mg iron/day. After 1 year of rhuEPO therapy, 9 out of 14 patients had oral iron supplementation. Serum ferritin decreased continuously during the first 3 months. Throughout the following study period it returned to the initial range (table 2).

Discussion

Increasing the hematocrit level during acute rhuEPO therapy requires a massive transfer of iron storage tissue to red blood cells to manufacture new hemoglobin. Iron-deficient erythropoiesis occurs when transferrin saturation

Table 2. Iron status during rhuEPO treatment in patients undergoing CAPD

	Before rhuEPO	On rhuEPO treatment, months				
		1	3	6	9	12
Oral iron supplementation (40–80 mg/day)						
patients, n	6	9	9	8	10	9
Ferritin, µg/l	332 ± 55	193 ± 45*	125 ± 44*	214 ± 74	179 ± 59*	327 ± 134

Mean values ± SEM. *p < 0.05; n = 14.

(serum iron divided by iron binding capacity and multiplied by 100) falls below 16% [18]. An increased red cell protoporphyrin is found at this level of transferrin saturation in patients with iron deficiency [19]. In order to detect iron deficiency before iron deficiency anemia arose, van Wyck et al. [20] used serum ferritin, serum iron, transferrin saturation, and serum iron increment after deferoxamine iron challenge as measures of body iron reserves. Projected iron needs for new hemoglobin synthesis were calculated by the equation:

Iron needs (mg) = $150 \times (Hb_{target} - Hb_{initial})$.

Calculations were found to be 100% sensitive, 71% specific and 93% efficient. Using the pre-rhuEPO hemoglobin and ferritin concentration, the net iron surplus or deficit after achieving target hematocrit can be predicted according to a nomogram [18, 20]. Of 22 patients where predicted iron needs exceed calculated available iron reserves, 20 developed iron deficiency before reacting target hemoglobin [20]. When the ferritin level falls below 50 mg/dl, or transferrin saturation falls below 20%, gastrointestinal iron absorption may be too slow to meet iron demands. In these cases parenteral iron therapy is recommended [18]. Nonazotemic patients with established iron deficiency respond as well to oral as to parenteral iron therapy [21]. MacDougall et al. [22], however, reported on 5 patients who responded poorly to rhuEPO therapy until they were given iron intravenously.

Blumberg et al. [23] compared in 20 patients undergoing continuous ambulatory peritoneal dialysis (CAPD) serum ferritin concentration with bone marrow iron stores (graded semiquantitatively). A close correlation between the two parameters was found. It was concluded that serum ferritin concentrations adequately reflect bone marrow iron stores and are useful as a guide to iron replacement therapy in patients undergoing CAPD [23]. On the other hand, Ali et al. [24] found that serum ferritin levels correlated well with the degree of hepatosplenic siderosis in hemodialysis patients treated with elemental iron given intravenously as iron dextran. However, serum ferritin did not always correlate with bone marrow iron stores in these patients. The authors found raised serum ferritin (mean 1,336 ng/ml) concentrations in 10 marrow iron-depleted hemodialysis patients [24].

In a recent study performed by Brunati et al. [25], serum ferritin and erythrocyte ferritin were evaluated in 35 patients on chronic hemodialysis treatment. Twenty-five of these patients with basal serum ferritin <500 µg/l were treated orally with 200 mg of Fe^{2+} for 2 months. Twenty patients did not respond to the therapy. In these patients, the mean serum ferritin increased significantly after therapy, while the mean erythrocyte ferritin levels did not change. Five patients responded to the therapy. In these 5 patients the mean serum ferritin did not change significantly after therapy, while erythrocyte ferritin increased significantly. The authors concluded that erythrocyte ferritin measurement could be an important and useful test in detecting the presence of an iron deficiency erythropoiesis in chronic hemodialysis patients [25].

It has been shown that red cell ferritin content reflects changes occurring in tissues both in iron deficiency and iron overload [26]. The relationship between the needs of iron for basal erythropoiesis and an expanded erythroid marrow can be monitored by measuring red cell protoporphyrin [27]. The diagnostic usefulness of bone marrow hemosiderin, serum ferritin, transferrin saturation, mean corpuscular volume and red cell protoporphyrin were elevated in hemodialysis patients. Moreb et al. [28] found that serum ferritin was useful in identifying subjects with both increased and reduced iron stores, whereas transferrin saturation could only be used for indicating iron overload. Red cell protoporphyrin, however, was not useful in either case [28].

Hochberg et al. [29] measured serum immunoreactive EPO levels in patients with iron deficiency anemia and patients with rheumatoid arthritis. The regressive slope was significantly greater in patients with iron deficiency than in patients with rheumatoid arthritis [30] and also in HIV-infected patients indicating a particular relationship between hemoglobin and EPO in iron-deficient patients.

In the present study both serum ferritin and erythrocyte ferritin fell during rhuEPO treatment in hemodialysis patients. There was also a decrease of neutrophil lactoferrin in 9 out of these 12 patients. None of these patients was supplemented with iron according to the high initial serum ferritin values. It is quite possible that redistribution of iron from liver to blood results in an increased lactoferrin in these 3 patients. In contrast, 6 of 14 CAPD patients were supplemented orally with iron before rhuEPO therapy. The fall of serum ferritin during rhuEPO treatment could be reversed by increasing the iron supplementation from 6 to 10 out of the 14 patients. Hemoglobin increased from 7.7 ± 0.3 to 10.5 ± 0.6 g/dl and erythrocytes from 2.58 ± 0.10 to 3.55 ± 0.19 million cells/µl [31].

Iron Overload

Sixty-eight out of 333 patients in the USA multicenter clinical trial with rhuEPO had iron overload defined as a serum ferritin level greater than 1,000 ng/ml. After 6 months of rhuEPO therapy, the mean serum ferritin levels decreased 39% from $3,179 \pm 258$ to $1,949 \pm 213$ ng/ml [1].

Transfusional iron overload causes hepatomegaly, reactive hepatic fibrosis and cirrhosis, abnormal liver enzymes, splenomegaly, myocardial dysfunction, proximal myopathy, endocrinopathies, carbohydrate intolerance, arthritis, osteomalacia, and/or an increased susceptibility to infection due to enhanced bacterial growth and virulence due to altered phagocytosis [32–39]. Hakim et al. [37] found in 62 (41%) of 150 unselected hemodialysis patients serum ferritin levels > 2,000 ng/ml.

Iron overload dialysis patients tend to have macrocytic erythrocytes and during therapy with rhuEPO this macrocytosis becomes even more pronounced. Eschbach [40] suggests that this is due to the increase in reticulocytes in the peripheral circulation. If iron removal from the organism is warranted, rhuEPO therapy can be combined with venesection. Lazarus et al. [41] studied 5 iron overloaded patients during 9–14 months of rhuEPO treatment in combination with phlebotomy. Serum ferritin decreased from 8,412 to 3,007 ng/ml and liver density, measured by computer tomography, declined from 89 to 71 Hounsfield units.

Mossey et al. [39] reported on 45 hemodialysis patients with transfusional iron overload treated 3 times weekly with intravenous deferoxamine mesylate during the dialysis treatment. Significant reductions in liver iron content were documented using a γ-ray scattering technique. This decrease in liver iron content could not, however, be predicted by clinical parameters or serum ferritin [39].

Conclusion

Patients should have a serum ferritin of at least 100 ng/ml before starting rhuEPO therapy. Every dialysis patient who is not iron overloaded should be on oral or intravenous iron during rhuEPO treatment [42]. In iron overloaded patients rhuEPO therapy may improve functional disturbances of various organs and immune dysfunction due to reduction in iron content.

References

1 Eschbach, J.W.; Downing, M.R.; Egrie, J.C.; Browne, J.K.; Adamson, J.W.: USA multicenter clinical trial with recombinant human erythropoietin (Amgen). Results in hemodialysis patients. Contrib. Nephrol., vol. 76, pp. 160–165 (Karger, Basel 1989).
2 Winearls, C.G.; Oliver, D.O.; Pippard, M.J.; Reid, C.; Downing, M.R.; Cotes, P.M.: Effect of human eythropoietin derived from recombinant DNA on the anaemia of patients maintained by chronic haemodialysis. Lancet *ii:* 1175–1177 (1986).
3 Eschbach, J.W.; Egrie, J.C.; Downing, M.R.; Browne, J.K.; Adamson, J.W.: Correction of the anemia of end-stage renal disease with recombinant human erythropoietin. New Engl. J. Med. *316:* 73–78 (1987).
4 Nonnast-Daniel, B.; Creutzig, A.; Kühn, K.; Bahlmann, J.; Reimers, E.; Brunkhorst, R.; Caspary, L.; Koch, K.M.: Effect of treatment with recombinant human erythropoietin on peripheral hemodynamics and oxygenation. Contrib. Nephrol., vol. 66, pp. 185–194 (Karger, Basel 1988).
5 Mohini, R.: Clinical efficacy of recombinant human erythropoietin in hemodialysis patients. Semin. Nephrol. *9:* (suppl. 1): 16–21 (1989).
6 Eschbach, J.W.; Adamson, J.W.: Recombinant human erythropoietin. Am. J. Kidney Dis. *11:* 203–209 (1988).
7 Zins, B.; Drüeke, T.; Zingraff, J.; Berheri, L.; Kreis, H.; Naret, C.; Delons, S.; Castaigne, J.P.; Peterlongo, F.; Casadevall, N.; Varet, B.: Erythropoietin treatment in anaemic patients on haemodialysis. Lancet *ii:* 1329 (1986).
8 Zehnter, E.; Pollok, M.; Ziegenhagen, D.; Bramsiepe, P.; Longere, F.; Baldamus, C.A.; Wellner, U.; Waters, W.: Urea kinetics in patients on regular dialysis treatment with recombinant human erythropoietin. Contrib. Nephrol., vol. 66, pp. 149–155 (Karger, Basel 1988).
9 Bommer, J.; Alexiou, U.; Müller-Bühl, E.; Eifert, J.; Ritz, E.: Recombinant human erythropoietin therapy in haemodialysis patients – dose determination and clinical experience. Nephrol. Dial. Transplant *2:* 238–242 (1987).
10 Stutz, B.; Rhyner, K.; Vögtli, J.; Binswanger, U.: Erfolgreiche Behandlung der Anämie bei Hämodialyse-Patienten mit rekombinantem humanem Erythropoietin. Schweiz. Med. Wochenschr. *117:* 1397–1402 (1987).
11 Schaefer, R.M.; Kürner, B.; Zech, M.; Krahn, R.; Heidland, A.: Therapie der renalen Anämie mit rekombinantem humanem Erythropoietin. Dtsch. Med. Wochenschr. *113:* 125–129 (1988).

12 Casati, S.; Passerini, P.; Campise, M.R.; Graziani, G.; Cesana, B.; Perisic, M.; Ponticelli, C.: Benefits and risks of protracted treatment with human recombinant erythropoietin in patients having haemodialysis. Br. Med. J. *295:* 1017–1020 (1987).

13 Akizawa, T.; Koshikawa, S.; Takaku, F.; Urabe, A.; Akiyama, N.; Mimura, N.; Otsubo, O.; Nikei, H.; Suzuki, Y.; Kawaguchi, Y.; Ota, K.; Kubo, K.; Marumo, F.; Maeda, T.: Clinical effect of recombinant human erythropoietin on anemia associated with chronic renal failure. A multi-institutional study in Japan. Int. J. Artif. Organs *11:* 343–350 (1988).

14 Urabe, A.; Takaku, F.; Mizoguchi, H.; Kubo, K.; Ota, K.; Shimizu, N.; Tanaka, K.; Mimura, N.; Nihei, H.; Koshikawa, S.; Akizawa, T.; Akiyama, N.; Otsubo, O.; Kawaguchi, J.; Maeda, T.: Effect of recombinant human erythropoietin on the anemia of chronic renal failure. Int. J. Cell. Cloning *6:* 179–191 (1988).

15 Bodemann, H.H.; Rieger, A.; Bross, K.J.; Schröter-Urban, H.; Löhr, G.W.: Erythrocyte and plasma ferritin in normal subjects, blood donors and iron deficiency anemia patients. Blut *48:* 131–137 (1984).

16 Rautenberg, W.; Neumann, S.; Gunzer, G.; Lang, H.; Jochum, M.; Fritz, H.: Quantitation of human lactoferrin as an inflammation marker by enzyme-linked immunosorbent assay (ELISA). Fresenius Z. Anal. Chem. *324:* 364 (1986).

17 Harbeck, R.J.; Hoffmann, A.A.; Redecker, S.; Biundo, T.; Kurnick, J.: The isolation and functional activity of polymorphonuclear leukocytes and lymphocytes separated from whole blood on a single percoll density gradient. Clin. Immunol. Immunopathol. *23:* 682–690 (1982).

18 Van Wyck, D.B.: Iron deficiency in patients with dialysis-associated anemia during erythropoietin replacement therapy: Strategies for assessment and management. Semin. Nephrol. *9* (suppl 2): 21–24 (1989).

19 Cook, J.D.; Finch, C.A.; Smith, N.J.: Evaluation of the iron status of a population. Blood *48:* 449 (1976).

20 Van Wyck, D.B.; Stivelmann, J.C.; Ruiz, J.: Iron status in patients receiving erythropoietin for dialysis-associated anemia. Kidney Int. *35:* 712–716 (1989).

21 Crosby, W.H.: The rationale for treating iron deficiency anemia. Arch. Intern. Med. *144:* 471–472 (1984).

22 MacDougall, I.C.; Hutton, R.D.; Cavill, I.; Coles, G.A.; Williams, J.D.: Poor response to treatment of renal anaemia with erythropoietin corrected by iron given intravenously. Br. Med. J. *299:* 157–158 (1989).

23 Blumberg, A.B.; Marti, H.R.M.; Graber, C.G.: Serum ferritin and bone marrow iron in patients undergoing continuous ambulatory peritoneal dialysis. JAMA *250:* 3317–3319 (1983).

24 Ali, M.; Rigolosi, R.; Fayemi, A.O.; Braun, E.V.; Frascino, J.; Singer, R.: Failure of serum ferritin levels to predict bone-marrow iron content after intravenous iron-dextran therapy. Lancet *i:* 652–655 (1982).

25 Brunati, C.; Piperno, A.; Guastoni, C.; Perrino, M.L.; Civati, G.; Teatini, U.; Perego, A.; Fiorelli, G.; Minetti, L.: Erythrocyte ferritin in patients on chronic hemodialysis treatment. Nephron *54:* 219–223 (1990).

26 Cazzola, M.; Bergamaschi, G.; Barosi, G.; Bellotti, V.; Caldera, D.; Civiello, M.M.; Quaglini, S.; Arosio, P.; Ascari, E.: Biologic and clinical significance of red cell ferritin. Blood *62:* 1078–1087 (1983).

27 Finch, C.A.: Erythropoiesis, erythropoietin, and iron. Blood *60:* 1241–1246 (1982).

28 Moreb, J.; Popovtzer, M.M.; Friedlaender, M.M.; Konijn, A.M.; Herschko, C.: Evaluation of iron status in patients on chronic hemodialysis: relative usefulness of bone marrow hemosiderin, serum ferritin, transferrin saturation, mean corpuscular volume and red cell protoporphyrin. Nephron *35:* 196–200 (1983).
29 Hochberg, M.C.; Arnold, C.M.; Hogans, B.B.; Spivak, J.L.: Serum immunoreactive erythropoietin in rheumatoid arthritis: impaired response to anemia. Arthritis Rheum. *31:* 1318–1321 (1988).
30 Spivak, J.L.; Barnes, D.C.; Fuchs, E.; Quinn, T.C.: Serum immunoreactive erythropoietin in HIV-infected patients. JAMA *261:* 3104–3107 (1989).
31 Steinhauer, H.B.; Lubrich-Birkner, I.; Schollmeyer, P.: Effect of human recombinant erythropoietin on dialysis efficiency in continuous ambulatrory peritoneal dialysis. Contrib. Nephrol., vol. 89 (Karger, Basel, in press).
32 Van Asbeck, B.S.; Verbrugh, H.A.; Van Oost, B.A.; Marx, J.J.M.; Imhof, J.W.; Verhoef, J.: *Listeria monocytogenes* meningitis and decreased phagocytosis associated with iron overload. Br. Med. J. *284:* 542–544 (1982).
33 Waterlot, Y.; Cantinieaux, B.; Hariga-Muller, C.; De Maertelaere-Laurent, E.; Vanherweghem, J.L.; Fondu, P.: Impaired phagocytic activity of neutrophils in patients receiving haemodialysis: the critical role of iron overload. Br. Med. J. *291:* 501–504 (1985).
34 Boelaert, J.R.; van Landuyt, H.W.; Valcke, Y.J.; et al.: The role of iron overload in *Yersinia enterocolitica* and *Yersinia pseudotuberculosis* bacteremia in hemodialysis patients. J. Infect. Dis. *156:* 384–387 (1987).
35 Flament, J.; Goldman, M.; Waterlot, Y.; Dupont, E.; Wybran, J.; Vanderweghem, J.I.: Impairment of phagocyte oxidative metabolism in hemodialyzed patients with iron overload. Clin. Nephrol. *25:* 227–230 (1986).
36 Tielemans, C.L.; Lenclud, C.M.; Wens, R.; Collart, F.E.; Dratwa, M.: Critical role of iron overload in the increased susceptibility of haemodialysis patients to bacterial infections. Beneficial effects of desferoxamine. Nephrol. Dial. Transplant *4:* 883–887 (1989).
37 Hakim, R.M.; Stivelman, J.C.; Schulman, G.; Fosburg, M.; Wolfe, L.; Imber, M.J.; Lazarus, J.M.: Iron overload and mobilization in long-term hemodialysis patients. Am. J. Kidney Dis. *10:* 293–299 (1987).
38 Winchester, J.F.: Management of iron overload in dialysis patients. Semin. Nephrol. *6* (suppl 1): 22–26 (1986).
39 Mossey, R.T.; Wielopolski, L.; Bellucci, A.G.; Wilkes, B.M.; Chandra, M.: Reduction in liver iron in hemodialysis patients with transfusional iron overload by deferoxamine mesylate. Am. J. Kidney Dis. *12:* 40–44 (1988).
40 Eschbach, J.W.: Discussion; in Koch, K.M.; Kühn, K.; Nonnast-Daniel, B.; Scigalla, P. (eds.): Treatment of Renal Anemia with Recombinant Human Erythropoietin, pp. 135–138 (Karger, Basel 1988).
41 Lazarus, J.M.; Hakim, R.M.; Nowell, J.; Maiga, M.: Recombinant human erythropoietin and phlebotomy in the treatment of iron overload in chronic hemodialysis patients (abstract). Kidney Int. *35:* 253 (1989).
42 Eschbach, J.W.: Panel Discussion; in Baldomus, C.A.; Scigalla, P.; Wieczorek, L.; Koch, K.M. (eds.): Erythropoietin: From Molecular Structure to Clinical Application, pp. 212–218 (Karger, Basel 1989).

Prof. Dr. Dr. W.H. Hörl, Medizinische Universitätsklinik,
Hugstetter Strasse 55, D-7800 Freiburg i.Br. (FRG)

Aluminium Interference in the Treatment with Recombinant Human Erythropoietin

Stefano Casati, Mariarosaria Campise, Claudio Ponticelli

Divisione di Nefrologia e Dialisi, Ospedale Maggiore Policlinico, Milano, Italia

The defective production of erythropoietin by diseased kidneys is the leading cause of anaemia in patients with chronic renal failure. However, other factors can also be involved in the pathogenesis of this complication. Unidentified uraemic toxins and iron deficiency are two well-recognized cofactors of uraemic anaemia. Parathyroid hormone (PTH) has also been claimed to have a toxic effect on bone marrow activity, but its role has probably been overestimated. More recently, some trace metals have also acquired the connotation of toxins for patients with end-stage renal disease [1, 2]. Among them, much emphasis has been given to the abnormal aluminium (Al) levels found in different uraemic syndromes such as dialysis encephalopathy [3], osteomalacic bone disease [4] and anaemia [5–8]. Historical observations reported development of anaemia only in case of very high serum levels of Al [2, 4]. More recently, however, it has been demonstrated that anaemia can develop also in the presence of moderately increased Al serum levels and that correction of hyperaluminaemia with deferoxamine (DFO) can improve anaemia [9, 10]. This would suggest that Al, even at low serum concentrations, may have toxic effects on bone marrow. Alternatively, it has been speculated that DFO may chelate unknown substances, different from Al, which would also inhibit erythropoiesis [9, 10].

Mechanisms of Aluminium-Induced Anaemia

Among the different mechanisms through which Al can induce anaemia, direct inhibition of the haem synthetic enzymes ferrochelatase [8, 11], uroporphyrin decarboxylase [12] and δ-aminolaevulinic acid dehydratase [13] have been proposed. In addition, in vitro studies have reported a possible direct interference of Al on iron metabolism [14]. It has been observed, in fact, that besides albumin [15], Al utilizes transferrin as a carrier, occupying at least one of the two iron sites on it [14].

Does Al produce a toxic effect per se or does it need complementary substances to explicate its action? Al-citrate per se, even at very high concentrations, does not inhibit the growth of human erythroid cells cultured, while the addition of transferrin to the milieu results in a marked, dose-dependent, inhibition of cell growth [16]. Interestingly, only erythroid colonies but not myeloid colonies growth was inhibited. On the other hand, Al-involved inhibition is inversely correlated to the percent saturation of transferrin [16]. When transferred to a clinical setting, these data might indicate that iron deficiency enhances the development of an Al-induced anaemia and might also explain why patients with high iron deposits show reduced gastrointestinal Al absorption [17].

It should be underlined that, in vivo, DFO is not able to release Al from transferrin [18], while it can release it from its binding sites within the mineralized bone [19]. Abreo et al. [20] studied this phenomenon in vitro, using Friend erythroleukemia cells (FEC) cultured in regular media to which transferrin alone, Al-citrate or Al-transferrin had been added. These authors found a linear increase in intracellular Al concentrations in FEC grown in media containing transferrin-bound Al. The Al accumulation in FEC resulted in an inhibition of the haemoglobin (Hb) synthesis and of cell growth. By contrast, iron and transferrin uptake in Al-loaded FEC were always significantly higher compared to control FEC, as a result of an increased transferrin receptor expression [20]. It seems therefore that Al exerts its intracellular toxic effect gaining access into the cells bound to transferrin. As the uptake of iron was not reduced, it can be hypothesized that the toxic effect of Al on the cells is distal to iron uptake processes [20]. Lee et al. [21], however, found an altered iron uptake by the erythroid cells in rats.

Interference of Aluminium with rhuEPO Treatment

The above-described mechanisms might also interfere with the response to recombinant human erythropoietin (rhuEPO) in dialysis patients. Conflicting conclusions have been reported however. In a study on 6 patients the response to rhuEPO was not related to erythroid refractoriness, shorter cell survival, PTH and Al excess [22]. Another study showed that the higher the haemoglobin level after rhuEPO, the higher the serum Al levels [23]. However, several studies found that the response to rhuEPO was inversely correlated with the degree of Al intoxication [24, 25]. Moreover, other papers reported that bone marrow activity is significantly depressed in patients with high Al levels independently of the levels of PTH [25, 26]. It seems therefore that Al can actually impair the erythropoietic response to rhuEPO in dialysis patients.

Handling Aluminium-Dependent Resistance to rhuEPO

Al-dependent resistance to rhuEPO can be handled by giving greater doses of rhuEPO. In our previous experience we could obtain full correction of anaemia by giving greater doses of rhuEPO [24]. Another way for overcoming resistance to rhuEPO is DFO. A recent study [27] showed that 4 patients who had only a partial response (Hb increase less that 2 g/dl) to rhuEPO given at doses much higher than the usual had very high serum levels. In these patients a full correction of anaemia could be obtained by reducing the Al levels with DFO. In our own experience [28], patients with higher Al levels required significantly greater doses of rhuEPO to reach the target haemoglobin levels. However, when the Al body pool was reduced by treating these patients with DFO, the rhuEPO requirements significantly lowered [28] (fig. 1). These clinical observations have been confirmed by an interesting experimental paper on rats [29]. A group of rats was given intraperitoneal Al, while another group was not; rhuEPO was given subcutaneously to both groups of animals. In the control group, haemoglobin levels increased significantly, while they decreased in the Al-intoxicated rats. In the second part of the study, some rats were first intoxicated with intraperitoneal Al. Then, after a short interval, rhuEPO was given subcutaneously to them and to Al-nonintoxicated rats. Anaemia ensued when Al was given intraperitoneally and was corrected when rhuEPO was given. The correction of anaemia, however, did not reach the same degree observed in control rats. It should be remarked that these animals were not uraemic and were acutely intoxicated with extremely high Al doses.

DFO, given generally at the end of haemodialysis to get a more complete chelation of Al [30], is effective in reducing the Al levels in the majority of the uraemic patients and consequently can improve the Al-induced anaemia [31]. Unfortunately, a difficult chelatability of Al from some organs or the impossibility to avoid the prescription of Al-containing phosphate binders

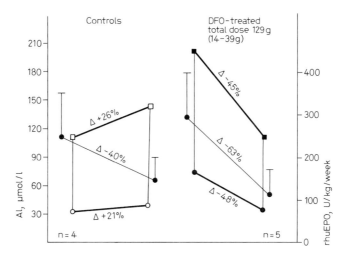

Fig. 1. Reversibility of the Al-induced resistance to rhuEPO.

make some patients particularly 'unresponsive' to DFO treatment. In these patients the Al body pool cannot be reduced and then rhuEPO doses cannot be tapered (fig. 2).

Correction of Cofactors Leading to Anaemia

Whatever the pathogenetic mechanism might be, our weapons to avoid anaemia in the uraemic patients, whether treated with rhuEPO or not, are the correct iron supplementation and the careful evaluation of patients' Al levels. There are today well-planned nomograms to calculate iron needs and iron supplementations [32]. Instead, there is still a wide disagreement on how Al pool should be clinically assessed. Some authors have demonstrated that baseline serum Al levels correlate with Al levels after DFO challenge [33]. Others stress the necessity of a DFO test [34, 35]. This test would be of particular value both in the case of acute or recent Al intoxication in which Al pool might be overestimated by the measurement of serum Al levels alone, and in the case of long-lasting chronic intoxication, where the Al pool might be underestimated. Other authors observed a great incidence of false-positive and false-negative results when only serum Al, either with or without

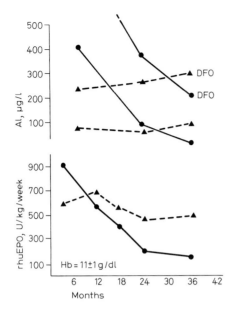

Fig. 2. rhuEPO dose requirements and Al levels.

DFO test, is considered. These authors claim a predominant role for the bone biopsy [36].

In our opinion it is advisable, although not always sufficient [16], to perform a DFO challenge [34] in order to have more complete data to evaluate the Al body pool. This noninvasive test, performed in all patients susceptible of receiving rhuEPO, allows to investigate patients' Al body pool, to approximately preview their sensibility to rhuEPO, and to treat with DFO all the patients with high Al levels. The lower the Al body pool, the higher should be the patient's sensibility to rhuEPO, thus allowing a sparing of rhuEPO, which is a very expensive drug.

The role of Al in inducing and maintaining anaemia in the uraemic patient has been established. It has not been completely defined, however, how Al toxicity exerts its action. Although inhibition of the haem synthesis [20] appears to be dependent upon transferrin [16], it seems that an altered iron uptake might also be involved [21].

Conclusions

Correction of iron deficiency and of Al intoxication are crucial to improve anaemia and to allow a full activity of rhuEPO. We have not yet clarified the role of Al on rhuEPO that another question arises: Can rhuEPO itself rise serum Al levels? In rhuEPO-treated patients, Loibl et al. [23] described a rise of the serum levels of Al, consequent to its increased mobilization, to gastrointestinal resorption or to interference of Al with iron metabolism. Further studies are needed to answer this very important question.

References

1 Drüeke T: Dialysis osteomalacia and aluminum intoxication. Nephron 1980;26: 207–210.
2 Parkinson IS, Ward MK, Kerr DNS: Dialysis encephalopathy, bone disease, and anaemia: the aluminum intoxication syndrome during dialysis. J Clin Pathol 1981; 34:1285–1294.
3 Alfrey AC, Mishell JM, Burks J, et al: Syndrome of dyspraxia and multifocal seizures associated with chronic haemodialysis. Trans ASAIO 1972;18:257–261.
4 Ward MK, Feest TG, Ellis HA, Parkinson IS, Kerr DNS, Herrington J, Goode G: Osteomalacia dialysis osteodystrophy. Evidence for a water borne aetiological agent, probably aluminium. Lancet 1978;i:841–845.
5 Eliott HL, MacDougall AI: Aluminium studies in dialysis encephalopathy. Proc EDTA 1978;15:157–163.
6 Touam M, Martinez F, Lacour B, Bourdon R, Zingraff J, DiGiulio S, Drüeke T: Aluminum-induced, reversible microcytic anaemia in chronic renal failure: clinical and experimental data. Clin Nephrol 1983;19:295–298.
7 Kaiser L, Schwartz K: Aluminum-induced anemia. Am J Kidney Dis 1985;6:348–352.
8 McGonigle RJS, Parsons V: Aluminum-induced anemia in hemodialysis patients. Nephron 1985;39:1–9.
9 De la Serna FJ, Gilsanz F, Ruilope LM, Praga M, Rodicio JL, Alcazar JM: Improvement in the erythropoiesis of chronic dialysis patients with desferrioxamine. Lancet 1988;i:1009–1012.
10 Altmann P, Marsh F, Plowman D, Cunningham J: Aluminium chelation therapy in dialysis patients: evidence for inhibition of haemoglobin synthesis by low levels of aluminium. Lancet 1988;i:1012–1015.
11 Huber C, Frieden E: The inhibition of ferroxidase by trivalent and other metal ions. J Biol Chem 1970;245:3970–3982.
12 Day RS, Eales L, Disler PB: Porphyrins and the kidney. Nephron 1981;28:261–267.
13 Meredith P, Moore P, Goldberg A: The effects of aluminum, lead and zinc on delta-aminolevulinic acid dehydratase. Enzyme 1977;22:22–27.
14 Trapp GA: Plasma aluminum is bound to transferrin. Life Sci 1983;33:311–316.

15 King SW, Savory J, Wills MR: Aluminum distribution in serum following hemodialysis. J Lab Clin Sci 1982;12:143–149.
16 Mladenovic J: Aluminium inhibits erythropoiesis in vitro. J Clin Invest 1988;81:1661–1665.
17 Cannata JB, Suarez-Suarez C, Cuesta V, Rodriguez Roza R, Allende MT, Herrera H, Perez Llanderal J: Gastrointestinal aluminum absorption: Is it modulated by the iron absorptive mechanism? Proc EDTA 1984;21:354–359.
18 Day JP: Chemical aspects of aluminium chelation by desferrioxamine; in Taylor A (ed): Aluminium and Other Trace Elements in Renal Disease. London, Baillière Tindall, 1986, pp 184–192.
19 Ackrill P, Day JP, Garstang FM, et al: Treatment of fracturing renal osteodystrophy by desferrioxamine. Proc EDTA 1982;19:203–207.
20 Abreo K, Glass J, Sella M: Aluminum inhibits hemoglobin synthesis but enhances iron uptake in Friend erythroleukemia cells. Kidney Int 1990;37:677–681.
21 Lee C, Van Wick D, Stivelman JC: Effect of short-term aluminum loading on ferrokinetics in rats. Am Soc Nephrol 1989;285(abstract book).
22 Adamson JW, Egrie JC, Haley NR, Schneider GL, Sherrard DJ, Eschbach JW: Why do some hemodialysis patients need large doses of recombinant erythropoietin? Am Soc Nephrol 1989;47(abstract book).
23 Loibl U, Meisl F, Stockenhuber F, Manker W, Meisinger V: Influence of erythropoietin therapy on aluminum metabolism in chronic hemodialysis patients. Am Soc Nephrol 1989;119(abstract book).
24 Casati S, Passerini P, Campise M, Graziani G, Cesana B, Perisic M, Ponticelli C: Benefits and risks of protracted treatment with human recombinant erythropoietin in patients having haemodialysis. Br Med J 1987;295:1017–1020.
25 Grützmacher P, Ehmer B, Messinger D, Scigalla P: Effect of aluminium overload and hyperparathyroidism on bone marrow response to rhuEPO therapy. Proc EDTA 1989;201(abstract book).
26 Hollomby DJ, Muirhead N, Hodsman AB, Cordy PE, Cordy PE, Slaughter D: The role of aluminum and PTH in erythropoietin resistance in hemodialysis patients. Am Soc Nephrol 1989;113(abstract book).
27 Gronhagen-Riska C, Honkanen E, Rosenlöf K, Tikkanen I, Fyhrquist F: Resistance to erythropoietin treatment is reversed by desferrioxamine; in De Broe ME (ed): Aluminum and Iron Overload in Haemodialysis. Toronto, Hogrefe & Huber, 1989, pp 56–61.
28 Casati S, Castelnovo C, Campise M, Ponticelli C: Aluminium interference in the treatment of haemodialysis patients with recombinant human erythropoietin. Nephrol Dial Transplant (in press).
29 Losekann A, Urena P, Khiraoui F, Casadevall N, Brigitte Z, Bererhi L, Zingraff J, Bourdon R, Drüeke T: Aluminium intoxication in the rat induces partial resistance to the effect of recombinant human erythropoietin. Nephrol Dial Transplant (in press).
30 Allain P, Chaleil D, Mauras Y, Beaudeau G, Varin MC, Poignet JL, Ciancioni C, Ang KS, Cam G, Simon P: Pharmacokinetics of desferrioxamine and of its iron and aluminum chelates in patients on haemodialysis. Clin Chim Acta 1987;170:331–338.
31 Ackrill P, Ralston AJ, Day JL, Hodge KC: Successful removal of aluminium from patients with dialysis encephalopathy. Lancet 1980;ii:692–693.

32 Van Wick DB, Stivelman JC, Ruiz J, Kirlin LF, Katz MA, Ogden DA: Iron status in patients receiving erythropoietin for dialysis-associated anaemia. Kidney Int 1989; 35:712–716.
33 DeVernejoul MC, Marchais S, London G, Bielakoff J, Chappuis P, Morieux C, Llach F: Deferoxamine test and bone disease in dialysis patients with mild aluminum accumulation. Am J Kidney Dis 1989;2:124–130.
34 Simon P, Allain P, Ang KS, Cam G, Mauras Y: Prophylaxie et traitement de l'intoxication aluminique chez l'insuffisant rénal chronique; in Crosnier J, Funck-Brentano JL, Bach JF, Grünfeld JP (eds): Actualités Néphrologiques de l'Hôpital Necker. Paris, Flammarion, 1984, pp 383–414.
35 Hodsman AB, Hood SA, Brown P, Condy PE: Do serum aluminum levels reflect underlying skeletal aluminum accumulation and bone histology before or after chelation by deferoxamine? J Lab Clin Med 1985;106:674–681.
36 Malluche HH, Slatopolsky E, Faugere MC: The deferoxamine infusion test: a multicentre study. Am Soc Nephrol 1989;121(abstract book).

Stefano Casati, MD, Divisione di Nefrologia e Dialisi,
Ospedale Maggiore di Milano, Via Commenda 15, I-20122 Milano (Italy)

Effect of Recombinant Human Erythropoietin on Anemia and Dialysis: Efficiency in Patients Undergoing CAPD

P. Schollmeyer, I. Lubrich-Birkner, H.B. Steinhauer

Department of Internal Medicine, Division of Nephrology, University of Freiburg i.Br., FRG

The first clinical studies already showed a dose-dependent increase in hematocrit in hemodialysis patients after application of erythropoietin (EPO) doses between 15 and 500 U/kg body weight [1, 2]. This dose was injected intravenously 3 times a week after hemodialysis. The typical initial dose of 3 × 50 U/kg body weight intravenously for hemodialysis patients results in a hematocrit increase from approximately 18% to 28–30% within 10–12 weeks. Intravenous application is appropriate in the case of hemodialysis patients, since there is vascular access and EPO can easily be injected intravenously after dialysis [3].

The intravenous route of application is not suitable for continuous ambulatory peritoneal dialysis (CAPD) patients since the patients usually see a doctor only every 4–6 weeks. Venous vascular puncture would require additional visits to the doctor. Therefore, several groups have investigated the possibility of subcutaneous or intraperitoneal EPO application for CAPD patients.

Intraperitoneal EPO Application – Pharmacokinetics

Animal experiments have shown a close linear relation between dialysate absorption and peritoneal EPO absorption, suggesting a convective transport of EPO [4]. This is supported by the fact that the molecular weight is more

Table 1. rhuEPO pharmacokinetics after i.v., i.p. and s.c. administration [data from 7]

	100 U rhuEPO/kg		
	i.v.	i.p.	s.c.
C_{max}, mU/ml	1,923 ± 197	213 ± 27	32 ± 4
t_{max}, h	0.3 ± 0	17.0 ± 2.3	28 ± 5
$t_{1/2}$, h	5.6 ± 0.3	8.0 ± 0.4	30.2 ± 5.3

than 30,000 daltons so that diffusion processes can only be of minor importance for peritoneal resorption.

50,000 U EPO in 2 liters 1.3% dialysate were administered to CAPD patients intraperitoneally. The retention time was 8 h [5]. This dose corresponds to 700 U/kg body weight. It is about 12 times higher than the typical initial dose. The highest serum EPO levels were obtained after 12 h (375 U/l). After 24 h, the levels amounted to 287 U/l. The bioavailability after intraperitoneal application with the dialysate was only 2.9%. However, the bioavailability was 6.8%, the half-life 11 h after intraperitoneal application of EPO immediately prior to 2 liters 1.5% dialysate, as compared to a half-life of 8.3 h after intravenous application [6]. The differences in the reabsorption of EPO via the peritoneum depend more on the filling level and peritoneal dialysis technique than on the dose. Krömer et al. [7] compared the pharmacokinetic data of 100 U/kg after intravenous, intraperitoneal and subcutaneous injections to 12 intermittent peritoneal dialysis (IPD) patients. The studies were carried out at weekly intervals. With intraperitoneal application, the dose was administered into the empty abdominal cavity at the end of IPD in order to prevent discharge with the dialysate effluent (the next IPD treatment was carried out 3 days later) (table 1). In spite of the difference in doses, the maximum concentrations were similar to the findings of MacDougall et al. [5]. Half-life was only slightly longer than with intravenous application. However, resorption was delayed and bioavailability amounted to 41%.

Within the framework of regular CAPD therapy, a bioavailability of only 3–7% may reasonably be expected. Due to the high costs and the danger of germ contamination, this type of application is not suitable for CAPD patients.

Subcutaneous EPO Application – Pharmacokinetics

Table 1 indicates that the lowest maximum serum concentrations are observed after subcutaneous application with slow resorption, as compared to intravenous or intraperitoneal application of the same doses. However, half-life is almost 6 times as long as after intravenous application. Different studies show a half-life between 15 and 30 h [5–8, 10]. Maximum serum levels of 79–89 U/l have been reached with subcutaneous injection of 40–100 U/kg body weight after 18–24 h [6, 8, 9]. As compared to intravenous injection (same group of patients), bioavailability amounted to 21.5% [10], whereas comparisons of different groups of patients yielded a result of 49% [6]. Subcutaneous application thus results in lower but longer lasting serum concentrations of EPO and a lower bioavailability than with intravenous injection.

Hematological Findings

An increase in hematocrit (Hct) can thus be obtained by means of intraperitoneal or subcutaneous EPO application, not only by means of intravenous application. The desired hematocrit value can be reached within 3 months by means of intraperitoneal application 3 times a week of the same dose as administered intravenously [9, 11, 12]. However, EPO is less easy to apply intraperitoneally, and the efficiency depends on the filling level of the peritoneal cavity and the amount of dialysate. Bommer et al. [13] have already proved the superior efficiency of EPO applied subcutaneously as compared to intravenous injection. They were able to maintain the hematocrit values previously obtained by means of intravenous injection with only half the dose applied subcutaneously for more than half a year.

In the meantime, the first results have been presented on subcutaneous EPO therapy for CAPD patients. With an application twice weekly of 30–60 U/kg body weight, it is possible to treat anemia to the desired extent [9, 10, 14, 15]. As compared to intraperitoneal application, the same dose injected subcutaneously resulted in a faster hematocrit and hemoglobin increase and the maintenance dose could be reduced to one third of the initial dose with subcutaneous application, as indicated in table 2 [12]. These findings demonstrate that subcutaneous EPO application is the easiest and most efficient way of administering the hormone to CAPD patients. In addition, subcutaneous injection is also the most cost-efficient type of EPO therapy.

Table 2. Comparison of efficiency of i.p. and s.c. rhuEPO administration: initial doses 100 U rhuEPO/kg 3 × weekly s.c. or i.p. in 1,000 ml dialysis fluid, dwell time 10–12 h [data from 12]

Weeks	EPO, U/kg		Hct, %	
	i.p.	s.c.	i.p.	s.c.
0	100	100	19	22
4	100	88	23	32
9	100	88	28	35
13	94	62	30	38
21	84	37	32	37

Table 3. rhuEPO long-term study in CAPD patients

CAPD patients	14 (10 f, 4 m), age 41.8 (21–65) years
CAPD treatment	25.4 (10–41) months
Hct	at beginning: 22.6%; target: 35%
Diet	K-low, protein-rich (>1.2 g/kg b.w.)
Controls	clinical status twice weekly; serum and dialysate chemistry once a week; ferritin, transferrin once a month
CAPD	according to clinical requirements one dialysate change at the outpatient department (1,500 ml, 1.36%, 4 h r.t.)
rhuEPO	50 U/kg s.c. twice weekly; increase of 25 U/kg, if Hct < 5%/month; 4-week interruption of therapy if Hct > 35%

Studies on Ultrafiltration and Peritoneal Clearance

Patients and Methods

Our studies are based on observations of 14 CAPD patients over a period of 12 months. Ten of these patients had to undergo dialysis due to chronic glomerulonephritis, 2 due to interstitial nephritis, 1 due to late diabetic syndrome and 1 due to cystic disease. Table 3 summarizes the initial situation and the course of the studies. At the beginning of the therapy, vitamin B_{12} and folic acid levels were within the normal range. Six patients with serum ferritin concentrations of less than 100 µg/l received iron supplementation with 40 mg/day. In addition to the hematological parameters, we studied peritoneal ultrafiltration and clearance. For this purpose, the patients underwent a supervised ambulatory dialysate exchange of 1.5 liters of the 3.6% dialysate with a 4-hour retention time. Antihypertensive therapy was free. All patients took calcium carbonate as a phosphate-binding medication. Only in case of phosphate values of more than 6 mg/dl, medication containing aluminium hydroxide was temporarily administered.

Table 4. Hematological data on rhuEPO treatment in CAPD patients

	Time, months					
	before rhuEPO	on rhuEPO treatment				
	−0.5	1	3	6	9	12
Hemoglobin, g/l	77 ±3	85 ±2	108* ±5	108* ±4	111* ±7	105* ±6
Hematocrit, %	22.6 ±3	26.2 ±4	30.8* ±3	33.1* ±3	34.2* ±4	33.4* ±4
Erythrocytes, 10^9/l	2.58 ±0.10	2.89 ±0.08	3.69* ±0.14	3.74* ±0.14	3.74* ±0.22	3.55* ±0.19
Leukocytes, 10^9/l	6.04 ±0.63	5.12 ±0.57	5.31 ±0.28	6.12 ±0.60	5.62 ±0.55	6.26 ±0.54
Platelets, 10^9/l	298 ±27	305 ±23	258 ±20	274 ±23	296 ±44	276 ±34

x̄ ± SEM, n = 14, *p < 0.03.

Results

Table 4 shows the development of the hematological parameters during the 12 months. The mean EPO dose of initially 50 U/kg body weight was increased to a mean dose of 67 U/kg body weight within the first 3 months and finally amounted to 48 U/kg body weight.

Subcutaneous injection of EPO was tolerated without significant adverse effects. Seven patients reported local pains which ceased after a few minutes. Four patients required more extensive hypertensive therapy. One patient described an influenza-like syndrome without rise in body temperature during the first 2 weeks of EPO treatment. The symptoms disappeared as the treatment progressed.

During the first 3 months a continuous rise in hematocrit values was obtained. This value remained largely stable throughout the remaining study time. Hemoglobin and the number of erythrocytes increased, while leukocyte and platelet counts indicated no changes. During the first 3 months, the mean serum ferritin levels decreased continuously to values of 125 ± 44 µg/l. In the further course of the study, these values increased again to reach the original range. In 3 patients, ferritin concentration fell below 20 µg/l,

Table 5. Peritoneal ultrafiltration (UF) and solute clearances (Cl) on rhuEPO treatment in 14 CAPD patients

	Time, months					
	run-in period	rhuEPO treatment				
	−0.5	1	3	6	9	12
UF, ml/h	41.8 ±6.2	49.5* ±9.4	57.3** ±9.6	54.0* ±5.8	46.8* ±7.8	55.8* ±11.0
Cl creatinine, ml/min/1.73 m²	4.11 ±0.23	4.41 ±0.28	5.14* ±0.38	4.35 ±0.27	4.39 ±0.24	4.49 ±0.35
Cl urea, ml/min/1.73 m²	5.81 ±0.31	5.80 ±0.26	6.64* ±0.45	5.83 ±0.26	5.74 ±0.27	6.00 ±0.38
Cl phosphate, ml/min/1.73 m²	3.44 ±0.17	3.99 ±0.29	4.56* ±0.30	4.18* ±0.38	3.96 ±0.34	4.38* ±0.52
Cl potassium, ml/min/1.73 m²	5.36 ±0.37	4.99 ±0.31	6.11* ±0.46	5.43 ±0.27	5.19 ±0.20	5.43 ±0.31

*p < 0.05; **p < 0.03.

indicating iron deficiency. The oral iron supplementation was doubled to 80 mg iron. Serum transferrin levels did not change significantly.

No changes were observed in the clinical parameters. During the 12 months, body weight rose by 300 g, which corresponds to the typical weight development under CAPD. Mean arterial blood pressure and heart rate did not change significantly during the study. Serum potassium (initial 4.84 ± 0.16 mmol/l) and phosphate (initial 1.49 ± 0.10 mmol/l) increased slightly but continuously during the first 6 months (potassium 5.38 ± 0.27 mmol/l, phosphate 1.84 ± 1.4 mmol/l; $p < 0.05$) and then decreased again. At the end of the study period, these values did not differ significantly from the initial values. Serum sodium, serum calcium, serum creatinine and serum urea remained unchanged.

Peritoneal ultrafiltration (1.5 liters dialysate, 1.36% glucose solution, 4-hour retention time) increased during the first 3 months of treatment from 41.8 ± 6.2 to 57.3 ± 9.6 ml/h. This increase remained constant throughout the rest of the study (table 5). Increased peritoneal ultrafiltration also resulted in increased creatinine, urea, phosphate and potassium clearances. Increased peritoneal ultrafiltration during EPO treatment corresponded positively to increased hematocrit values.

Discussion

We have recently reported on the positive effects of short-term subcutaneous EPO treatment on renal anemia in CAPD patients and improved ultrafiltration [14]. The findings obtained during the 1 year treatment of 14 patients confirm those results.

Subcutaneous EPO injection (50–100 U/kg body weight) twice weekly resulted in a continuous rise in hematocrit from 23 to 35% during the first 4 months. With a maintenance dose of 48 U/kg body weight administered twice weekly, this hematocrit value could be kept constant throughout the rest of the study. As opposed to the findings of other groups, we did not observe increases in the number of platelets, which is typical of intravenous EPO treatment of hemodialysis patients [16, 17]. The leukocyte counts did not show significant changes either.

The EPO-induced increase in hematocrit corresponded positively to improved peritoneal ultrafiltration. This observation seems to be particularly important, since a long-term treatment of chronic kidney insufficiency through CAPD may very well go along with decreased ultrafiltration and clearance rates [18, 19]. Several studies report increased peripheral vascular resistance after correction of anemia by means of EPO [20, 21]. This was attributed to decreased compensatory hypoxic vasodilation after improved tissue oxygenation. No information is available on the effects of anemia on peritoneal vascular resistance. It seems possible that an EPO-induced increase in vascular resistance may cause a blood redistribution in terms of enhanced mesenteric blood flow. Since increased mesenteric blood flow results in improved transperitoneal transport [22], we may reasonably conclude that the enhanced EPO-induced ultrafiltration and the higher clearance rates are caused by improved perfusion of the peritoneal capillary bed.

Hormonal changes such as suppression of basal plasma renin activity and plasma aldosterone concentration as well as increased atrial natriuretic peptide might be attributed to EPO-induced hypervolemia as a result of enhanced erythropoiesis [23].

In accordance with other studies on the effect of EPO [24, 25], all CAPD patients reported improved physical well-being and improved general well-being after correction of anemia. A rise in the retention values in serum that has been observed in hemodialysis patients [1, 2] did not occur in our patients during CAPD treatment. Only the serum potassium and serum phosphate values rose slightly during the first months of treatment. This may be attributed to dietary changes, since 12 of 14 patients reported increased appetite. However, after 12 months of treatment the serum electrolyte values and the weights did not differ significantly from the initial values. The increased arterial blood pressure in 4 originally hypertensive patients could be easily managed by means of intensified antihypertensive therapy. After 12 months of treatment, the mean arterial pressures did not differ significantly from the initial values. We did not observe higher blood pressures in patients who were normotensive prior to EPO treatment. Successful EPO treatment requires that the iron levels be within the normal range [26]. Three patients with ferritin levels of less than 50 µg/l at the beginning of the treatment developed signs of iron deficiency (ferritin below 20 µg/l) during the first 3 months. This necessitated increased iron supplementation. As opposed to the findings in hemodialysis patients [2, 26, 27], intravenous iron application was not necessary during EPO treatment. Oral supplementation of 40–80 mg Fe^{2+} was sufficient. After 12 months of treatment and oral iron supplementation, the serum ferritin levels were within the upper normal range.

To summarize, subcutaneous EPO treatment of anemic CAPD patients is an efficient and safe method. After 1 year of treatment, the maintenance dose was 48 ± 2 U/kg body weight twice weekly. This dose was sufficient to keep hematocrit within the desired range of 32–35%. No major side effects could be observed. The lasting improvement of ultrafiltration after partial correction of anemia by means of EPO should be able to considerably improve liquid balance and CAPD efficiency.

References

1 Winerarls CG, Oliver DO, Pippard MJ, Reid OL, Downing MR, Cotes PM: Effect of human erythropoietin derived from recombinant DNA on the anemia of patients maintained by chronic haemodialysis. Lancet 1986;ii:1175–1179.
2 Eschbach JW, Ergie JC, Downing MR, Browne JK, Adamson JW: Correction of the anemia of end-stage renal disease with recombinant human erythropoietin. New Engl J Med 1987;316:73–78.
3 Fisher JW, Bommer J, Eschbach J, Fried W, Lange RD, Massry S, Nathan D, Nolph KD, Teehan BP: Statement of the clinical use of recombinant erythropoietin in anemia of end-stage renal disease. NKF position paper. Am J Kidney Dis 1989;14: 163–169.

4 Bargman JM, Breborowcz A, Rodela H, Sombolos K, Oreopoulos DG: Intraperitoneal administration of recombinant human erythropoietin in uremic animals. Periton Dial Int 1988;8:249–252.
5 MacDougall JC, Roberts DE, Neubert P, Dharmasena AD, Coles GA, Williams JD: Pharmacokinetics of recombinant human erythropoietin in patients on continuous ambulatory peritoneal dialysis. Lancet 1989;i:425–427.
6 Kampf D, Kahl A, Passlick J, Pustelnik A, Eckardt KU, Elmer B, Jacob C, Baumelon A, Grabensee B, Gahl GM: Single dose kinetics of recombinant human erythropoietin after intravenous, subcutaneous and intraperitoneal administration. Contrib Nephrol. Basel, Karger, 1989, vol 76, pp 106–111.
7 Krömer G, Solf A, Elmer B, Kaufmann B, Quellhorst E: Single dose pharmacokinetics of recombinant human erythropoietin comparing intravenous, subcutaneous and intraperitoneal administration in IPD patients (abstract). Kidney Int 1990;37:331.
8 Neumayer HH, Brockmöller J, Fritschka E, Roots J, Saigalla P, Wattenberg M: Pharmacokinetics of recombinant human erythropoietin after s.c. administration and long-term i.v. treatment in patients on maintenance hemodialysis. Contrib Nephrol. Basel, Karger, 1989, vol 76, pp 131–142.
9 Lui SF, Chung WWM, Leung CB, Chang Lai KN: Pharmacokinetics and pharmacodynamics of subcutaneous and intraperitoneal administration of recombinant human erythropoietin in patients on continuous ambulatory peritoneal dialysis. Clin Nephrol 1990;33:47–51.
10 MacDougall IC, Cavill J, Davies ME, Hutton RD, Coles GA, Williams JD: Subcutaneous recombinant erythropoietin in the treatment of renal anaemia in CAPD patients. Contrib Nephrol. Basel, Karger, 1989, vol 76, pp 219–226.
11 Frenken LAM, Coppens PJW, Tiggeler RG, Struijk DG, Krediet RT, Koene RAP: Clinical study of the efficacy and tolerance of recombinant human erythropoietin after intraperitoneal administration in patients treated with continuous ambulatory peritoneal dialysis. Kidney Int 1989;36:508.
12 Coppens PJW, Frenken LAM, Struijk DG, Tiggeler RGWL, Krediet RT, Koene RAP: Comparative study of intraperitoneal and subcutaneous administration of recombinant human erythropoietin in patients treated with CAPD (abstract). Kidney Int 1990;37:326.
13 Bommer J, Ritz E, Weinreich T, Brommer G, Ziegler T: Subcutaneous erythropoietin. Lancet 1988;ii:406.
14 Steinhauer HB, Lubrich-Birkner I, Dreyling KW, Hörl WH, Schollmeyer P: Increased ultrafiltration after erythropoietin-induced correction of renal anemia in patients on continuous ambulatory peritoneal dialysis. Nephron 1989;53:91–92.
15 Pirano B, Johnston JR: The use of subcutaneous erythropoietin in CAPD patients. Clin Nephrol 1990;33:200–202.
16 Bommer J, Alexiou C, Müller-Bühl U, Eifer J, Ritz E: Recombinant human erythropoietin therapy in hemodialysis patients – dose determination and clinical experience. Nephrol Dial Transplant 1987;2:238–242.
17 Grützmacher P, Bergmann M, Weinreich T, Nattermann U, Reimers E, Pollok M: Beneficial and adverse effects of correction of maintenance hemodialysis. Contrib Nephrol. Basel, Karger, 1988, vol 66, pp 104–113.
18 Slingeneyer A, Canaud B, Mion C: Permanent loss of ultrafiltration capacity of the peritoneum in long-term peritoneal dialysis: an epidemiological study. Nephron 1983;33:133–138.

19 Wideröe T-E, Smeby LC, Mjaland S, Dahl K, Berg KJ, Aas TW: Long-term changes in transperitoneal water transport during continuous ambulatory peritoneal dialysis. Nephron 1984;38:238–247.
20 Nonnast-Daniel B, Creutzig A, Kühn K, Bahlmann J, Reimers E, Brunkhorst R, Caspary L, Koch KM: Effect of treatment with recombinant human erythropoietin on peripheral hemodynamics and oxygenation. Contrib Nephrol. Basel, Karger, 1988, vol 66, pp 185–194.
21 Steffen HM, Brunner R, Müller R, Degenhardt S, Pollok M, Lang R, Baldamus CA: Peripheral hemodynamics, blood viscosity, and the renin-angiotensin system in hemodialysis patients under therapy with recombinant human erythropoietin. Contrib Nephrol. Basel, Karger, 1989, vol 76, pp 292–298.
22 Maher JF: Pharmacologic manipulation of peritoneal transport; in Nolph KD (ed): Peritoneal Dialysis. The Hague, Nijhoff, 1981, pp 213–239.
23 Kokot F, Wiecek A, Grzeszczak W, Klepacka J, Klin M, Lao M: Endocrine abnormalities in patients with end-stage renal failure; in Hörl WH, Schollmeyer P (eds): New Perspectives in Hemodialysis, Peritoneal Dialysis, Arteriovenous Hemofiltration and Plasmapheresis. Adv Exp Biol 1989;260:61–68.
24 Mayer G, Thum J, Cada EM, Stummvoll HK, Graf H: Working capacity is increased following recombinant human erythropoietin treatment. Kidney Int 1988;34:525–528.
25 Lunding AP: Quality of life: subjective and objective improvements with recombinant human erythropoietin therapy. Semin Nephrol 1989;9:22–29.
26 Van Wyck DB, Stivelman JC, Ruiz J, Kirlin LF, Katz MA, Ogden DA: Iron status in patients receiving erythropoietin for dialysis-associated anemia. Kidney Int 1989;35:712–716.
27 Eschbach JW, Downing MR, Egrie JC, Browne JK, Adamson JW: USA multicenter clinical trial with recombinant human erythropoietin (Amgen). Results in hemodialysis patients. Contrib Nephrol. Basel, Karger, 1989, vol 76, pp 160–165.

Prof. P. Schollmeyer, Medizinische Universitäts-Klinik, Abteilung IV, Nephrologie, Hugstetterstrasse 55, D-7800 Freiberg i.Br. (FRG)

Does Treatment of Predialysis Patients with Recombinant Human Erythropoietin Compromise Renal Function?

Robert A.P. Koene, Leon A.M. Frenken[1]

Department of Medicine, Division of Nephrology, University Hospital, Nijmegen, The Netherlands

Recombinant human erythropoietin (rhuEPO) has proved to be an effective drug in the treatment of anaemia in patients with end-stage renal failure undergoing chronic haemodialysis or peritoneal dialysis. In many patients with chronic renal failure not yet on dialysis, anaemia starts to develop when the creatinine clearance has decreased below 25 ml/min. With the further progression of renal failure, signs and symptoms of anaemia appear that are not different from those in dialysis patients. It can therefore be expected that treatment with rhuEPO will also be of benefit to anaemic predialysis patients as it is for the dialysis-dependent population.

The most important question that needs to be answered is whether correction of the haematocrit to almost normal values in these patients will be hazardous to the kidney and will increase the rate of progression of renal failure. Experiments in animals have demonstrated that an acute increase of the haematocrit causes a rise in glomerular capillary pressure [1]. As pointed out by Brenner et al. [2], a constantly elevated glomerular capillary pressure may be responsible for the most often unrelenting progressive course of renal failure by causing focal glomerulosclerosis. A few studies in rats treated with rhuEPO have indeed shown that such a treatment may accelerate the development of renal failure [3, 4]. Before rhuEPO can be prescribed routinely to anaemic patients with chronic renal failure not yet on dialysis, it

---

[1] We thank Drs. R. Verberckmoes, P. Michielsen, H. Sluiter, and G. Schrijver for expert help and advice during the clinical trial. Recombinant human erythropoietin was provided by Cilag BV (Herentals, Belgium).

is thus of utmost importance to establish whether a damaging effect will also occur in the clinical situation.

We have studied the efficacy and safety of rhuEPO treatment in predialysis patients during correction of the anaemia and during maintenance therapy. Special attention was paid to changes in renal function and renal haemodynamics. Together with a review of the available literature on predialysis patients, this will be the subject of the current report.

Patients and Methods

Twenty-four patients were included in the trial: 13 females and 11 males, aged 23–68 years. They all had a known history of progressive chronic renal failure, whereas other clinically significant diseases were absent. Endogenous creatinine clearances ranged from 5 to 22 ml/min. All patients were anaemic with haemoglobin ranging from 5.3 to 10.2 g/dl and haematocrit from 0.16 to 0.30 l/l. The anaemia could not be attributed to other causes. Hypertension was either absent or medically controlled. The patients received stable doses of oral iron supplementation (up to 200 mg of elemental iron per day) and folic acid for 2 weeks before study entry and for the duration of the study. The patients were randomly assigned to three dosage groups each containing 8 patients. The doses were 50, 100 and 150 units rhuEPO/kg body weight (U/kg) per injection. After a 2-week period of baseline measurements, the patients were given intravenous injections of rhuEPO 3 times a week for 8 weeks. At the completion of this 8-week study, or whenever a patient's haematocrit exceeded the target values (37% for females, 39% for males) by two percentage points, the patient was entered into a maintenance study. During the maintenance phase rhuEPO was injected intravenously once a week. The starting dose was 3 times the dose given per injection during the correction phase, adjusted for response. Haemoglobin, haematocrit, and blood cell counts were determined twice weekly in the correction phase and at least once monthly in the maintenance phase. Blood pressure (supine, before venepuncture) was monitored before each dosing.

The serum creatinine values of the individual patients during the 2 years preceding the study were retrieved from the clinical records. The reciprocal of serum creatinine can be used to estimate the rate of deterioration of the glomerular filtration rate (GFR). We used the equivalent 1,000 divided by serum creatinine (µmol/l) to calculate this rate. When sufficient values of serum creatinine recorded before therapy were available, linear regression was applied to the reciprocal of serum creatinine versus time (days) data. In 8 patients we were able to determine haemodynamic parameters before and after 8 weeks of treatment with rhuEPO. Effective renal plasma flow (ERPF) and GFR were determined using standard renal clearance techniques with continuous infusion of *p*-aminohippurate and inulin.

Differences from baseline levels within groups were evaluated using Student's t test for paired data. The slopes of the regression lines obtained before and after the start of rhuEPO therapy were compared with analysis of covariance. Comparison of several groups of data was done with analysis of variance. A p value of less than 0.05 was considered significant. Unless otherwise stated, all values are expressed as means ± standard deviation.

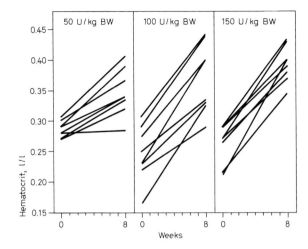

Fig. 1. Effect of three different dosages of rhuEPO on the haematocrit levels in predialysis patients (correction phase).

Results

Treatment with rhuEPO increased haematocrit in all dosage groups (fig. 1). An increase in haematocrit was observed in all patients, except one in the 50 U/kg dosage group. The rate of increase in haematocrit was significantly smaller in the 50 U/kg dosage group, whereas rates were comparable in the two other dosage groups. Anaemia was corrected during the 8-week study period in 87.5% of patients in the 150 U/kg dosage group, and in 50% of patients in the two lower dosage groups. This difference was not significant. In the two highest dosage groups an increase in reticulocyte counts was seen within 1 week of the start of the treatment. In these groups there was also a small but significant increase in thrombocyte counts. However, the increase was transient and values never exceeded the upper limit of the normal range.

The mean subjective scores for ability-to-do-work and energy level increased in all dosage groups during rhuEPO treatment.

During the correction phase there were no significant changes in mean arterial pressure (MAP) in either of the three dose groups. Also, mean systolic and diastolic blood pressure and MAP did not change during the maintenance phase. Additional data from the maintenance phase revealed that the mean number of antihypertensive drugs prescribed increased significantly from 1.5 ± 1.1 before the start of rhuEPO treatment at 2.0 ± 1.4 after 10 weeks of therapy ($p < 0.05$). This was caused by the increase of antihypertensive medications in 9 of 18 previously hypertensive patients. The remaining 6 patients who were normotensive at the start of the therapy remained so during the whole trial period.

None of the 24 patients required haemodialysis during the first 8 weeks of the study period. The mean serum creatinine and creatinine clearances did not change significantly during the correction phase. Additional information on the evolution of renal function can be derived from the 6-month maintenance study that followed the correction phase. In 14 patients sufficient data could be obtained to compare the time course of renal function (as measured by 1,000/creatinine versus time) before and during rhuEPO therapy. The periods reviewed included 20 months before and 7 months during therapy. There were no significant differences between the slopes of 1,000/creatinine versus time before therapy (-0.036 ± 0.002) and during therapy (-0.031 ± 0.004), indicating that there was no significant change in the progression of renal failure during the observation period.

In addition to the measurements of serum creatinine, a special study on renal haemodynamics was done. ERPF and GFR were determined in 8 patients before and after 12 weeks of treatment with rhuEPO. The results of this study are shown in figure 2. There were no significant changes of these parameters after 12 weeks of treatment. Moreover, the filtration fraction (FF) and the fractional excretion of albumin did not change significantly.

Discussion

The results of this therapeutic trial in predialysis patients demonstrate that rhuEPO treatment can effectively correct anaemia in these patients. This is in accord with the results obtained by three other groups that have so far published their experience [5–7]. A summary of all studies is given in table 1.

The haematologic response does not appear to differ from that in patients with end-stage renal failure who are treated by chronic dialysis. In almost all cases in which a response was absent this could be attributed to intercurrent illnesses or early withdrawal from the trials. All studies show that treatment with rhuEPO improves the patients' subjective ratings of well-being, ability to do work, and energy level. The exact dose requirements for rhuEPO in predialysis patients cannot be simply derived from these studies. There may be considerable differences between individual patients. A dose of 50 U/kg, 3 times a week intravenously, seems to be the preferable approach

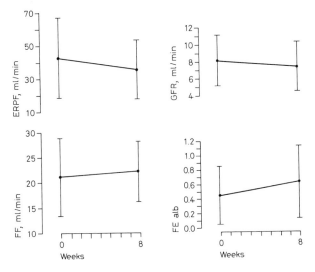

Fig. 2. Renal haemodynamic studies in 8 patients before and after 8 weeks of treatment with rhuEPO.

Table 1. Summary of haematologic responses during the correction phase

Reference	Patients, n	Duration weeks[1]	Haematocrit	
			start	end
Stone et al. [5]	8	8	0.32	0.40
Lim et al. [6]	11	8	0.27	0.38
Eschbach et al. [7]	17	8–12	0.27	0.37
This study	24	8	0.26	0.37

[1] Or until the target haematocrit was reached.

during the correction phase. In the study that included a maintenance phase of 6 months the dose requirements decreased. This may, at least partly, be related to the change from intravenous to subcutaneous administration. There is preliminary evidence that during subcutaneous administration, dose requirements are about 30% less than during intravenous administration [8]. More experience is needed before the exact dosing schedule and the optimal route of administration can definitively be established.

A rise in blood pressure during treatment with rhuEPO has been observed in both normotensive and hypertensive dialysis patients. Since regulation of volume balance differs between haemodialysis patients and predialysis patients, it is not self-evident that the latter patients will show a similar untoward blood pressure response to rhuEPO therapy. The experience in haemodialysis patients suggests that such changes, in case they occur, will become mostly manifest during the correction phase of the anaemia. In the study of Stone et al. [5] hypertension developed in 2 of 10 predialysis patients. In 1 patient it was transient whereas in the other patient accelerated hypertension occurred that made hospitalization necessary. Lim et al. [6] did not observe changes in mean systolic or diastolic blood pressure after 8 weeks of treatment in 11 predialysis patients. However, they increased antihypertensive medication in 3 patients. In the study of Eschbach et al. [7] additional antihypertensive medications were required in 9 of 14 previously hypertensive patients. Hypertension developed in 2 of 3 patients who were normotensive before therapy. In our study, hypertension developed in none of the normotensive patients, but in half of the previously hypertensive patients antihypertensive medications were increased during the study. The highest incidence of hypertensive events occurred among patients with the greatest rates of change in haematocrit. These results suggest that, without additional antihypertensive treatment, blood pressure will rise in predialysis patients. Careful monitoring of blood pressure and adjustment of the antihypertensive regimen, especially during the correction phase of the anaemia, are therefore indicated when predialysis patients are treated with rhuEPO. Serious problems with blood pressure can probably largely be prevented by a relatively slow correction of the anaemia, because this gives more time for haemodynamic adaptation and it provides better opportunity for timely adjustment of the antihypertensive medication.

In experimental studies in rhuEPO-treated rats that underwent five-sixths nephrectomy to induce renal failure, the progression of the disease was accelerated when compared to untreated five-sixths nephrectomized controls [3, 4]. Garcia et al. [4] demonstrated that in these rats an increase in intraglomerular capillary pressure developed, which the investigators thought to be a consequence of an increased vascular resistance due to the rise in whole blood viscosity. These observations have raised concern about the use of rhuEPO in predialysis patients. However, one should realize that in the rat experiments there was also a considerable increase in systemic blood pressure during rhuEPO treatment. It seems likely that this hypertension played an important role in accelerating the progression of renal failure in these animals, since hypertension is known to be important in determining the rate of development of renal damage in this model.

In the studies with rhuEPO in predialysis patients, all investigators have tried to monitor blood pressure meticulously and to adjust antihypertensive treatment when necessary. So far, the experience suggests that under these conditions detrimental effects on renal function as observed in rats do not occur in man. In the studies of Lim et al. [6] and of Eschbach et al. [7] creatinine clearances did not decrease significantly after 8 weeks of rhuEPO treatment. During the correction phase of our study, there was also no significant decrease in creatinine clearances. Our renal haemodynamic studies confirmed the conclusions derived from the creatinine clearance data. Therefore, it can be safely concluded that a relatively rapid correction of the anaemia in 8–12 weeks does not have acute detrimental effects on renal function. With regard to the effects of long-term correction of the anaemia on renal function, the available data do not permit definitive conclusions. Eschbach et al. [7] compared the slopes of the reciprocal serum creatinine values versus time curves during rhuEPO therapy (median follow-up 12 months) with the pretreatment slopes in 17 patients, and found no significant change.

In our study there was no significant change in the rate of progression of renal failure in 14 patients during 7 months of rhuEPO therapy. However, small changes in the rate of progression will be difficult to detect. Ideally, a placebo-controlled trial with a prolonged observation period in a large group of predialysis patients would be necessary to detect such changes. However, it is highly questionable whether a placebo-treated control group can be recruited. A possible alternative will be to study a large group of predialysis patients in whom pretreatment follow-up is long enough to permit calculations of the rate of progression of their renal failure and compare these with the progression rate during one or several years of rhuEPO treatment.

The results of these first studies in anaemic predialysis patients demonstrate that this group will also benefit from treatment with rhuEPO. Long-term studies are required to assess the effect of this treatment on renal function. On the basis of the current information it can be expected that a possible detrimental effect on renal function will not be very large. Moreover, the benefits of treatment with rhuEPO will probably outweigh a possible adverse effect on renal function. Therefore, predialysis patients seem to be eligible for treatment with rhuEPO, provided that blood pressure is carefully monitored and controlled.

References

1 Myers BD, Deen W, Robertson CR, Brenner BM: Dynamics of glomerular ultrafiltration in the rat. VIII. Effects of hematocrit. Circ Res 1975;36:425–435.
2 Brenner BM, Meyer T, Hostetter T: Dietary protein intake and the progressive nature of kidney disease. The role of hemodynamically mediated glomerular injury in the pathogenesis of progressive glomerular sclerosis in aging, renal ablation, and intrinsic renal disease. New Engl J Med 1982;307:652–659.
3 Gretz N, Lasserre JJ, Meisinger E, Strauch M, Waldherr R, Kraft K, Weidler A: Potential side-effects of erythropoietin. Lancet 1987;i:46.
4 Garcia DL, Anderson S, Rennke HG, Brenner BM: Anemia lessens and its prevention with recombinant erythropoietin worsens glomerular injury and hypertension in rats with reduced renal mass. Proc Natl Acad Sci USA 1988;85:6142–6146.
5 Stone WJ, Graber SE, Krantz SB, Dessypris EN, O'Neil VL, Olsen NJ, Pincus TP: Treatment of the anemia of predialysis patients with recombinant human erythropoietin: A randomized, placebo-controlled trial. Am J Med Sci 1988;296:171–179.
6 Lim VS, DeGowin RL, Zavala D, Kirchner PT, Abels R, Perry P, Fangman J: Recombinant human erythropoietin treatment in predialysis patients. Ann Intern Med 1989;110:108–114.
7 Eschbach JW, Kelly MR, Haley NR, Abels RI, Adamson JW: Treatment of the anemia of progressive renal failure with recombinant human erythropoietin. New Engl J Med 1989;321:158–163.
8 Bommer J, Ritz E, Weinreich T, Bommer G, Ziegler T: Subcutaneous erythropoietin. Lancet 1988;ii:406.

Leon A.M. Frenken, MD, Department of Medicine,
Division of Nephrology, University Hospital Nijmegen,
PO Box 9101, NL–6500 HB Nijmegen (The Netherlands)

Subject Index

Administration
  intraperitoneal 95
  intravenous 63
  subcutaneous 31, 63, 95, 109
Altitude 18
Aluminium, intoxication 60, 87
δ-Aminolevolinic acid dehydratase 88
Analgesic abuse 60
Anemia
  Blackfan-Diamond syndrome 33
  bone marrow transplantation 33
  Fanconi syndrome 32
  pure red cell anemia 32
  sickle cell 13
  thalassemia 13

Bioavailability 63, 96
Blood
  loss 59
  pressure 105, 107, 110
  viscosity 52

Cells
  B lymphocytes 40
  BFU-E 1
  CFU-GM 4
  CFU-Mix 4
  endothelial 17
  F 13
  macrophage 40
  monocytes 40
  peripheral blood
    mononuclear 39
    progenitor 1
  T helper 40
Chemotherapy 32

Chronic myelogenous leukemia 29
Clearance
  creatinin 111
  peritoneal 100
Costs 65
Cytokines 68

Deferoxamine 83, 87
Dialysis
  CAPD 17, 80, 95
  hemodialysis 36, 42, 59, 78
  IPD 96

Effective renal plasma flow 109
Endocrine system
  gastrin 47
  glucagon 46
  glucose tolerance 44
  insulin 45
  pancreas polypeptide 48
Erythrocyte(s)
  filterability 55
  life span 55, 88
  red cell
    ferritin 79
    protoporphyrin 81
  reticulocytes 4, 7, 12, 61
  volume 55
Erythropoietin
  AIDS 28, 72, 82
  chronic inflammation 72
  dose 61, 65
  gene 26
  half-life 96
  hepatitis 72
  iron deficiency 82
  malignancy 28, 72

Subject Index

Erythropoietin (cont.)
  mRNA 26, 70
  polycythemia 22
  production site 16, 27, 68
  receptor 27, 30, 40
  regulation 16
  renal failure 72
  rheumatoid arthritis 28, 82
  serum levels 18, 21
  transplantation 22

Ferritin 29, 79
Ferrochelatase 87
α-Fetoprotein 70
Filtration fraction 108

Glomerular capillary pressure 105
Glomerular filtration rate 109

Haem synthesis 87
Hematocrit 4, 12, 29, 53, 62, 99, 107
Hemoglobin
  A 13
  F 12
Hypersensitivity 36
Hypoxia 19, 74

Immunoglobulins 37
Interferon τ 72
Interleukin-1 70
Interleukin-3 7
Iron
  deficiency 78
  elimination therapy 83
  need 81
  overload 28, 78
  supplementation 80, 102

Kidney
  adult 27
  isolated perfused 16
  native 23
  transplanted 19

Lactoferrin
  neutrophil 79
  plasma 79
Liver
  computer tomography 83
  fetal 27

Macrocytosis 83
Mean corpuscular volume 4, 83
Megakaryocytes 30
Multiple myeloma 31
Myelodysplastic syndrome 32

Non-Hodgkin's lymphoma 28
Nucleotides 26

Oxygen
  saturation 16
  sensor 17
  supply 16
  tension 16

Parathyroid hormone 87
Phagocytosis 83
Pharmacokinetics 96
Phlebotomy 83
Placenta 40
Pokeweed mitogen 37
Polycythemia 22
Polytransfusion 60, 78
Predialysis 105
Progression 105

Sialoglycoprotein 26
Side-effects
  flu-like syndrome 3
  hematoma 63
  hypertension 3
  urticaria 63
Siderosis 82

Transferrin 88
  saturation 80
Transplantation 19
Thrombocytes 4, 107
Thyroid hormone 42
Tumor necrosis factor-α 33

Ultrafiltration 98
Uremic toxins 15, 48, 60
Uroporphyrin decarboxylase 88

Virulence 83